JAMES LAWRENCE POWELL is executiv
Consortium, and former director
Museum of Natural History. He t
College, where he also served as a
to the Cretaceous and *Mysteries of Terra Firma,* he lives in Buellton, California, with his wife, three cats, two dogs, and a burro.

"Reads like a detective story . . . Professional or amateur geologists and anyone else who has an interest in great rivers or has viewed the Grand Canyon will find this book captivating." —*Library Journal*

"Powell's well-crafted account makes one appreciate just how [the Grand Canyon] came to be so grand." —*Natural History*

"This gem of a book is both entertaining and educational . . . a great read for anyone traveling through the American Southwest." —*Escapees*

"*Grand Canyon* will appeal to those who have stared at the great chasm in amazement, and simply asked how it came to be." —*High Country News*

"*Grand Canyon* will be enjoyed by anyone who is curious about how geologists think, piece together disparate information, and assemble explanations." —*Science*

"What many readers will walk away with is a sense of the awesome power of water running over the surface of the earth." —*Kirkus Reviews*

"The Grand Canyon's beauty, grandeur, and striking form have made it one of the greatest tourist attractions in the U.S., and also one of the greatest intellectual challenges to geologists. James Powell's exciting account of the Canyon's development is worthy of the excitement that the Canyon itself inspires."
—Jared Diamond, Pulitzer Prize–winning author of
Guns, Germs, and Steel and *Collapse*

"An engaging and lucid account of one of geology's greatest monuments. The story of how the Colorado River cut the Grand Canyon turns out to be a remarkable detective story, complete with red herrings and innocent suspects. The table of the Grand Canyon encapsulates features of the growth in our knowledge over the whole of the earth sciences."
—Richard Fortey, author of *Earth* and *Trilobite!*

Grand Canyon

Solving Earth's
Grandest Puzzle

James Lawrence Powell

A PLUME BOOK

PLUME
Published by Penguin Group
Penguin Group (USA) Inc., 375 Hudson Street,
New York, New York 10014, USA
Penguin Group (Canada), 90 Eglinton Avenue East, Suite 700, Toronto, Ontario, Canada M4P 2Y3
(a division of Pearson Penguin Canada Inc.)
Penguin Books Ltd., 80 Strand, London WC2R 0RL, England
Penguin Ireland, 25 St. Stephen's Green, Dublin 2,
Ireland (a division of Penguin Books Ltd.)
Penguin Group (Australia), 250 Camberwell Road, Camberwell, Victoria 3124,
Australia (a division of Pearson Australia Group Pty. Ltd.)
Penguin Books India Pvt. Ltd., 11 Community Centre, Panchsheel Park,
New Delhi – 110 017, India
Penguin Group (NZ), cnr Airborne and Rosedale Roads, Albany,
Auckland 1310, New Zealand (a division of Pearson New Zealand Ltd.)
Penguin Books (South Africa) (Pty.) Ltd., 24 Sturdee Avenue,
Rosebank, Johannesburg 2196, South Africa

Penguin Books Ltd., Registered Offices: 80 Strand, London WC2R 0RL, England

Published by Plume, a member of Penguin Group (USA) Inc. Previously published in a Pi Press edition.

First Plume Printing, December 2006
1 3 5 7 9 10 8 6 4 2

℗ REGISTERED TRADEMARK—MARCA REGISTRADA

CIP data is available.
ISBN: 0-13-147989-X (hc.)
ISBN: 0-452-28787-1 (pbk.)

Printed in the United States of America
Original hardcover design by Kim Llewellyn

Contents

Photo section follows page 152.

Figure 1. *The Colorado River and the Colorado Plateau*

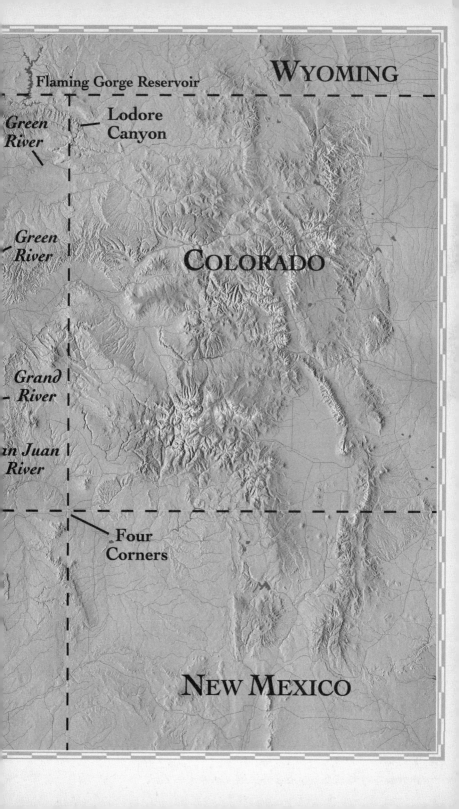

PART ONE

A Brief History of Awe

Chapter 1

Six Feet

In the year 1540, a squadron of men from Coronado's expedition, seeking the fabled Seven Cities of Gold, came instead upon the rim of a great canyon. Though their Indian guides said the stream at the bottom was half a league (about a mile) across, the Conquistadors estimated it at six feet. Three of the lightest and most agile men attempted to descend to the river, but after clambering down for most of a day, they got only about one-third of the way and gave up. Rock spires that from the rim had appeared no taller than a man turned out to be higher than the Great Tower of Seville. In 1604, the next expedition to arrive, led by Juan de Oñate, Governor of New Mexico, found a muddy, silt-laden river and named it el Rîo Colorado, "the Red River." The stream turned out to be the Little Colorado, not the main stem, the first of many confusions and disagreements over the name of the river. Most of two centuries would elapse before another Spanish explorer, the intrepid Father Garcés, passed by on his way to the Hopi villages. Offering neither gold nor souls, the country of the Big Cañon was too vast and barren for the Spaniards to comprehend. They saw no reason to remain or return.

The first American to see the cañon, Army Lieutenant Joseph Christmas Ives, agreed. He declared it "altogether valueless," adding that "Ours has been the first and will undoubtedly be the last, party of whites to visit the locality. It seems intended by nature that the Colorado River along the greater portion of its lonely and majestic way, shall be forever unvisited and undisturbed." Not until John Strong Newberry, the first geologist to arrive, did anyone begin to appreciate the Big Cañon. He wrote, "The Colorado plateau is to the geologist a paradise. Nowhere on the earth's surface, so far as we know, are the secrets of its structure so fully revealed as here." The next geologist on the scene, John Wesley Powell, concurred, saying, "The grand cañon of the Colorado will give the best geological section on the continent." The third, the dry, Euclidian Grove Karl Gilbert, said: "The Plateau province offers valuable matter in an advantageous manner"—for him, praise indeed. The scenery and geology inspired Clarence Dutton, the one Grand Canyon geologist who also qualified as a poet, to write, "It would be difficult to find anywhere else in the world a spot yielding so much subject matter for the contemplation of the geologist; certainly there is none situated in the midst of such dramatic and inspiring surroundings."

The four geologists were among the greatest of the nineteenth century, or, for that matter, any century. Collectively, they redefined the science of geology and gave it a distinctly American cast, while on the other side of the Atlantic, Charles Darwin was redefining biology and giving it a rather British complexion. The Colorado Plateau was a fount of scientific insight for these American scientists in the way the Galapagos Islands proved to be Darwin's land of inspiration. The plateau provided raw, variegated rock unobscured by vegetation or glacial drift; layer cake bedding with few faults and folds; incised canyons to provide the essential, but usually scarce, third dimension. The pioneer geologists had good reason to believe that they would soon decipher the geologic history of the Plateau and its rivers and learn lessons that would apply everywhere. How astounded they would have been to find that more than a century-and-a-half after Newberry's arrival, scientists still debated the history of the Colorado River and the origin of the Grand Canyon. Indeed, a few months into the twenty-first century, seventy-seven geologists spent a week on the rim of the Grand Canyon arguing their favorite subject all day and into the night. By the time the symposium ended, not only had they failed to reach consensus, but more theories than ever were on the table. Surprisingly, what had seemed to the pioneers to be an easy geologic puzzle to solve proved just the opposite.

Though the rock exposures of the Colorado Plateau are nakedly displayed and appear simple to understand, in fact the geologic history of the Colorado River and its canyons turns out to be deceptively complicated, vastly more so than the history of that ideal American river: the Mississippi. So fond are we of our longest and widest river that it has earned a set of respectful nicknames: Father of Waters, Old Man River, and Big River. In commerce, the Mississippi is our most important river; in music, history, and literature, our most celebrated. The Mississippi is the epitome of a river—the end to which, it is natural to think, all rivers head.

On its long journey, the Mississippi travels more or less directly south from the Twin Cities to salt water, there to deposit its sediment load and construct its elaborate birds-foot delta. The Mississippi not only is navigable its entire length, it is comprehensible. Already a sizable stream when it leaves the Minnesota lake country, the Mississippi grows steadily larger as other streams, some great in their own right, pay tribute. But the river never changes in any fundamental way. Why should it? It has no mountain ranges to avoid, no canyons into which to plunge, no quirks of geology to accommodate. The Mississippi can just keep rollin' along, leaving those seeking white-water adventure, or a river that can tell them more about how a continent and its rivers evolve, to look elsewhere. In flood it is dangerous and has cost many lives and billions of dollars in property damage, yet even then the Father of Waters remains like the streams with which we are familiar, only bigger, and, at those times, much more frightening.

Visitors to the Colorado River and the Grand Canyon recognize at once that they have come upon a "valley" and a river fundamentally different from the Mississippi and the other familiar streams of the eastern United States. The first glimpse of the giant chasm is such a shock that, years later, most of us can recapture the emotion it inspired. For some, we feel as we did at momentous times in our lives: the birth of a child, the death of a parent, the assassination of a president, or the end of a war. Ever after, we remember the instant when, traveling north through a beautiful pine forest, we arrived at Grand Canyon National Park and approached the rim. At first we could not see the Grand Canyon itself, for unlike a mountain range, or even the Mississippi River valley, the canyon remains almost entirely hidden until one is right on top of it. Then suddenly, without warning, the land fell away and there, where it had been all along, spread before us an unimaginably wide and deep chasm. Nothing we had ever seen—no other river, and no photograph or film of the Grand Canyon—truly prepared us for the sight. Years and decades later, we can bring back the sense of disbelief that we felt and remember how quickly it passed into awe. Those of us who are not expert photographers soon understood that our puny efforts and modest equipment could not do the scene justice. After a few perfunctory shots, we put away our cameras and simply gazed.

Inevitably, a few decide to hike down the Bright Angel or Kaibab trails to get a closer look. Like Coronado's men, most are in for a rude surprise. Just as the apparent simplicity of Grand Canyon geology tempted the pioneer geologists, so the easy downhill trip tempts a hiker to go farther and farther—why not at least to the edge of the Tonto Plateau? It is on beginning the return hike that the immensity of the Grand Canyon comes home. Still, by resting at strategic stops, determined not to require rescue by mule or helicopter, the hiker eventually regains the plateau, exhausted, proud, and with a new appreciation of the magnitude of the Grand Canyon.

After recovering from the shock of seeing the canyon for the first time, questions occur. Why is the Grand Canyon so different from other river valleys, even other western canyons? What caused it? Why is it located here and nowhere else? In a flash of inspiration, untold numbers of visitors have realized that the answer has become obvious to them! Some rare, terrible force ripped the earth's surface apart and provided a channel for the tiny stream a mile below. First came the chasm, then the river. But those who espouse this theory ought not forget the saying, "For every difficult question, there is an answer that is clear and simple and wrong."

The geologists of two hundred years ago, decades before any had seen the Grand Canyon, endorsed the simple answer. They believed that all valleys had been pre-created for the rivers that flow in them. These devout men accepted the Biblical account of earth history and thought that the turbulent waters of Noah's flood had excavated valleys, into which the water then naturally flowed. As late as the 1890s, even Clarence King, first Director of the U.S. Geological Survey, believed that valleys came before the rivers that lie in them. Today some claim that not only did the canyon come before the river, the entire history of the Grand Canyon—indeed, the entire history of the earth—compresses into only the last 10,000 years. But two centuries of progress have taught that, because rivers have the deep time of geology at their disposal, they can and do carve their own valleys.

The history of any valley or any canyon, no matter how grand, is inseparable from the history of the river that occupies it. The upper-most headwaters of the Colorado River lie in the Wind River Mountains of Wyoming, where its longest tributary, the Green River, rises. From there to the delta in the Gulf of California, the Green and Colorado together run for 1,750 miles. The Grand Canyon, impressive as are its depth and width, measures only about one-sixth of that total. The section of the river in the Grand Canyon is thus but one part of the whole, albeit the most important part. To understand the canyon, we have to first understand the river. The river is the parent, the canyon the child.

By the 1860s, Indians, explorers, mountain men, and Army survey-ors had traveled each of the major rivers in the United States, save one: the Colorado. Trappers knew the upper Green, the site of their annual rendezvous in the 1830s, perhaps better than they remembered their homes back east. Spaniards, Major Ives, and others had explored the Colorado River from its mouth in the Gulf of California past Fort Yuma and on up to the vicinity of present-day Las Vegas (see Figure 15, page 216 for the geography of the lower Colorado River). But in between the map was blank. At the border between Wyoming and Utah, the river disappeared through an ominous rock portal; a thousand miles down-stream, it debouched from a great canyon. What lay in between no one knew, for in that long stretch explorers had reached the Colorado River at only a handful of places. Even as late as the mid-1860s, not just the river, but the entire Colorado Plateau, as large as several eastern states combined, remained unexplored.

Those who did set out to explore the Colorado Plateau had no maps and almost no word of mouth. If any had gone before, they had not lived to tell the tale, much less bring back a map. Among the many inventions of the late twentieth century, none would have been of more use to the pioneer geologists and explorers than the map of the Colorado River and the Colorado Plateau, Figure 1. Whether the map would have reassured or terrified the pioneer explorers, it would certainly have revealed a river completely different from the Mississippi and its eastern kin.

Had the earliest plateau geologists had such a map, they could have traced the Green and Colorado Rivers all the way to the Gulf of California. Figure 1 and Figure 2 on page 15 track the river's journey starting in the alpine lakes of the Wind River Range, from whence the Green River descends across the high plains of Wyoming to the eponymous town on today's Interstate 80. From there it continues southward to the Utah-Wyoming border, where lies the Uinta Range, with peaks reaching 13,000 feet. In the nineteenth century, at this point the river entered a fabulous, vermilion-walled canyon, which Powell named the Flaming Gorge. Today that name is attached to the man-made lake that occupies the valley. Nearly alone among the mountains of the Western Hemisphere, the Uintas trend east-west, placing the range directly athwart the path of the southwarding river. The Green heads straight for the mountain range; then, as if losing its nerve, it swerves ninety degrees to the east and runs for fifty miles parallel to the foot of the mountains through an elongated natural park. By traveling only a few miles farther east, the Green could have avoided the Uintas altogether, but instead it suddenly swings back south and, through the menacing Gate of Lodore, enters a canyon incised into the heart of the mountains.

After twenty miles of white water, the Green exits Lodore and comes out into a valley only to enter another canyon, this one cut right through the center of the aptly named Split Mountain. From there the river passes through Desolation Canyon, then Gray Canyon, cut into the Book and Roan Cliffs, respectively. Downstream, the (formerly named) Grand River enters from the east, and the direction of the (now)

Colorado River swings a bit to the west. Cataract Canyon comes next, and after that, Glen Canyon, now submerged under Lake Powell. As the river crosses the Arizona line, below the Glen Canyon Dam, it begins to swing back south again, passing on its right the mouth of the Paria River and the plateau of the same name. It flows straight through Marble Canyon, heading for the Kaibab Plateau, the highest obstacle in its path after the Uinta Range. Figure 8, page 188, shows this section at a larger scale.

Although from the map it appears that the river could have bent less steeply by following a route south through the valley of the present-day Little Colorado River, that route is uphill. Instead, the river cuts across the nose of the Kaibab Plateau and then swings northwest. Thirty miles later, the Colorado makes another sharp bend back to the south-west, placing it on a collision course with the next high plateau, the Shivwits. But the river shifts direction and runs straight beside the Shiv-wits Plateau for thirty miles, then bends around it and resumes its northwesterly course. At the Grand Wash Cliffs, which we will meet repeatedly (see Figure 8, page 188 and Figure 11, page 199), the Colorado leaves the Grand Canyon and flows out into the Basin and Range province and modern Lake Mead. Here it turns ninety degrees south and now, finally, like the Mississippi, proceeds for several hundred miles with little deviation, traveling most of the way in a single large valley, until at last it reaches the Gulf of California (see Figure 15, page 216). There it has built a delta and today, if thirsty irrigators and desert cities had not impounded and extracted all its water, would continue to do so.

To a geologist, nothing beats a map. When maps of sea floor topography became available in the late 1940s, for the first time scientists could see the enormous faults; the gargantuan, world-encircling, under-sea mountain ranges; and the abyssal trenches that scar the seafloor. By the 1960s, to deny that something terrible had happened in the ocean basins, something on the scale of the continents themselves, had become impossible. Geologists could no longer dismiss as absurd the idea that continents had drifted: the seafloor appeared to retain the very tracks of their ponderous passage. In the same way, the modern map that we have been following reveals that the Colorado River has had anything but the simple history of the Mississippi. The Colorado crosses three distinct geologic terrains: the Colorado Plateau, Canyon Country, and the Basin and Range Province. It cuts deep canyons in some mountain ranges and goes around, or nearly around, others; changes direction unpredictably; meanders here and runs straight there; and generally behaves without apparent rhyme or reason.

The beauty of the Grand Canyon and the unparalleled exposure of its rocks have tempted generations of geologists. The unexpected complexity of the Canyon's geologic history confounded even the best of them, though gradually and collectively they have been able to put boundaries around the possible explanations. The pioneers had been sure that by studying the Colorado River and the Grand Canyon, they would learn lessons that eastern rivers could never teach them, lessons that would apply everywhere. They were right, but the actual lessons turned out to be quite different from those they expected. As we will see, those lessons have even deeper implications for earth history. One hundred and forty years of studying Grand Canyon geology have paid off, but in ways that have surprised even the best geologists.

The myth of the "scientific method" makes it seem that science is so logical—indeed, inevitable—that serendipity has no place. But in reality, it's just the opposite. And in geology, at least, some of the most important discoveries have come about serendipitously. And the most important of all were, strange to say, counterintuitive. Take one recent advance: when we observe the quiet night sky, we see the planets move in their predictable orbits and the moon pass through its familiar phases; save for the rare flash of a shooting star, nothing much seems to happen. Until the space age, we had no reason to suspect that our solar system was born in colossal, random violence and that since then, the impact of asteroids and comets has been its most fundamental process. Indeed, only in the last two decades have we discovered that *Homo sapiens* might exist only because, through a roll of cosmic dice, one errant comet or asteroid, out of the thousands that fly through space, happened to strike the Yucatan peninsula sixty-five million years ago and exterminate the dinosaurs and seventy percent of all living species.

In the same way, a rafter floating down the Colorado has no reason to suspect the power with which geologic time endows rivers. Who would guess that they engage in a vicious competition for territory that causes the more energetic streams to capture the waters of the less energetic? That only the fittest survive to carve deep canyons? The pioneer geologists of the Grand Canyon, outstanding as they were for their day, could not have imagined just how far the lessons of the Big Cañon would extend. They had no way to know that, by studying the imprisoned rivers, vast erosion, and dramatic uplift of the Colorado Plateau, they would uncover the underpinnings of a scientific revolution a century ahead. But such is the way of science.

Like a raft trip on the Colorado's white water, following the efforts of five generations of geologists to understand the river and its Grand Canyon stimulates our thinking, but, we soon recognize, requires more concentration and effort than does a riverboat cruise on the mighty Mississippi. The protagonist of this story had canoed the Father of Waters and other eastern streams for hundreds of miles. Yet those streams failed to teach him the lessons that the canyons of the Colorado brought home almost from the moment he launched onto their waters.

It is obvious that if the geologists of the nineteenth century had correctly understood the origin and history of the Grand Canyon, someone would have written a book like this one more than a century ago. The only remaining reason for geologists to hold a conference on the canyon in the year 2000 would have been for self-congratulation. Instead, the meeting was necessary because even today, geologists continue to try to answer what has turned out to be one of the most complicated questions they ever asked: what caused the Grand Canyon? Inevitably, in tracing the work of the many geologists who have attempted to answer the question, we are going to have to follow trails that at first look promising, but which in the end peter out. As the geologists have had to do, we will then back up and follow another trail. The obligation of the author is to mark the journey clearly and to play fair by not trumping up theories beyond what they deserve.

In only a few hundred years, science has made astounding progress. Geology and paleontology, two of its oldest disciplines, teamed at the start of the scientific revolution in the mid-1600s when a conflicted anatomist discovered a shark's tooth *inside* a rock. Paleontology later provided the evidence for Darwin's theory of evolution; geology eventually established the deep time that he required. As we will see, the trail from the Grand Canyon reaches all the way to the revolutionary theory of plate tectonics. And today, planetary scientists even search for evidence of plate tectonics on other bodies in the solar system. The pattern of observation, theory, experiment, and discovery continue as the magic carpet of science flies on.

Chapter 2

Water Catch 'Em

n May of 1869, two history-making events took place a few
hundred miles from each other, one in Utah, and the other in
Wyoming. On May 10, at Promontory, Utah, the Golden Spike
joined the Union Pacific rail line, coming from the east, and the Central
Pacific, coming from the west. The telegraph flashed "done" across the
country, the two coasts united, and the young nation achieved its great
dream. Now only four days of pleasant, unprecedentedly smooth riding
from San Francisco would carry a traveler all the way to Omaha. Only a
few more days would bring first Chicago and then New York. In the
decades before the Golden Spike, wagoneers on the Oregon Trail had
taken six months to cover only half the country. Soon after the two rail
lines met, the Transcontinental Express left New York and arrived in San
Francisco 83 hours and 39 minutes later.

Two weeks after the historic marriage of tracks, a great change had
taken place in the little Wyoming railroad town of Green River. As long
as it enjoyed the status of end-of-track to the westwarding Union Pacific,
serving as the latest location of Hell-on-Wheels, the burg's population
stood at a grand 2,000. But by May 24, 1869, the two sides of the
continent had joined and the very concept of end-of-track had vanished.
The town of Green River had nearly vanished as well, its denizens
having shrunk to a mere 100.

A traveler making one of the first transcontinental crossings, riding in relative opulence and at a speed of twenty miles per hour, gazing out the window in the early afternoon while passing through the little town, would have seen a strange sight. Ten men climbed into four wooden boats and launched them south down the Green River, from which the town took its name. Where in the world did they think they were going? The apparent leader of the crew was the least prepossessing of the lot—a wiry, bewhiskered fellow with no right arm, who barely reached 5' 6". Whatever the purpose of this odd flotilla, its captain seemed not to have won his position by virtue of his appearance or physical prowess (though subsequent events were to make one wonder what he could have done with two arms). Nor had his reputation preceded him: the youngest crew member wrote to his brother, "I suppose you never herd off him and he is a Bully fellow you bett." The fleeting glimpse the rail passenger might have caught of the little armada, its crew waving goodbye to the townsfolk assembled on the bank, would have been the last anyone was to hear of the party for thirty-seven days. And what they heard then was a bodacious lie.

Wallace Stegner, the poet laureate of the West, enumerates what the small navy was not:

It was not a government expedition. It had no imperial purposes. It was going to plant no flags. It was not motivated by fear that someone else would slip in and gain an advantage. It had no official backing, no congressional appropriation, [no sponsorship from] any department or bureau of government, nor any great semi-official organization, such as the Philosophical Society, which had backed Lewis and Clark. [Its] meager funds came out of the [leader's] own pocket, from the Illinois Natural History Society and from the Illinois Industrial University.

What was the purpose of such a poorly supported expedition? "To find out. To observe, analyze, map, comprehend, know. For the same reason the bear went over the mountain." Years afterwards, the leader said they had gone because "The Grand Canyon of the Colorado will give the best geological section on the continent." In other words, the journey had an academic, a scholarly, a scientific, but not a pecuniary,

Figure 2. *Green River Country*

purpose. One would surely expect that with such a goal, the crew would include credentialed scientific experts—professorial types who could describe and interpret the flora, the fauna, and the geology, as well as the Native Americans and their culture. Science was already advancing rapidly in the young nation and specialists were not hard to find; the other western geological surveys had them—as well as much else that the Green River expedition lacked. Its leader was the only member with any claim of college training. And though he bore the title of Professor of Geology at Illinois Wesleyan University, the education of John Wesley Powell (1834–1902) had come more from tramping through fields and canoeing rivers than from the rudimentary Midwestern colleges at which he lighted for a semester or two. Had he bothered to ask, he likely would have found few scientists willing to risk not only their lives, but what was almost as serious, their reputations, to join a young man of whom they had never heard on a voyage to a place no one had ever been.

The leader and his men also lacked any idea of the route the Green and Colorado rivers would follow. They knew little more than how they would be able to tell when their trip ended. Their destination was the mouth of the Virgin River, 1,000 miles downstream and 6,000 feet lower in elevation (see Figure 1). Powell and his men headed into the last great, unexplored region of the United States: the Colorado Plateau, a blank spot on the map the size of France or Germany. Others before him had discovered that two hundred miles or so downstream, the Grand River joined the Green to make the Colorado. In 1921, the Colorado Legislature renamed the Grand River the Colorado, the name previously reserved for the section downstream from the junction of the Green and Grand. This change was chauvinistic, confusing, and, since the Green travels a greater distance than the former Grand and is the true headwaters, flatly inaccurate.

From the merger of the Green and Grand, the Colorado ran south, then west through uncharted terrain, until, just above the mouth of the Virgin River, it debouched from a great canyon. From there, the river ran on through explored territory to meet the Gulf of California. Because the men knew the altitude of their launch point, the town of Green River, Wyoming, and the altitude of their destination, the Virgin

River, they knew that the river and anyone riding it had to drop more than a mile. But exactly how the river accomplished that descent no one knew. One did not have to be a geologist to predict that where canyons incise thousands of feet into solid rock, one should expect rapids and maybe worse. Even half a Niagara lurking unseen around a bend would terminate the ambitions and lives of the crew, likely without anyone ever knowing what had happened to them.

Because rowing upstream against the fast current was not an option, there were only two ways to escape after the party entered the canyons. One was to attempt to climb up and out. But the surrounding cliffs were often thousands of feet high; if and when one surmounted them, the nearest settlement was hundreds of miles away, across an uncharted desert. And as three of Powell's men were sadly to discover, the route out lay through what was still Indian country. The other option was to keep going, rapids and waterfalls permitting. Powell recognized that in these inaccessible canyons, a waterfall could be fatal in more than one way. In a deep, inaccessible canyon, a boatman did not actually have to raft over a waterfall for the experience to be fatal:

There are great descents yet to be made, but, if they are distributed in rapids and short falls, as they have been heretofore, we will be able to overcome them. But, maybe, we shall come to a fall in these canyons which we cannot pass, where the walls rise from the water's edge, so that we cannot land, and where the water is so swift that we cannot return. Such places have been found, except that the falls were not so great but that we could run them with safety. How will it be in the future!

Powell recalled a conversation he had the spring before the voyage with an old Indian named Pá-ri-ats, who told him about one of his tribe attempting to run the canyon. "'The rocks,' he said, his hands held above his head, his arms vertical, 'and the rocks h-e-a-p, h-e-a-p high; the water go h-oo-woogh, h-oo-woogh; water-pony (boat) h-e-a-p buck; water catch 'em; no see 'em Injun any more! no see 'em squaw any more! no see 'em papoose any more!'" But in spite of the jocularity, Powell and his men risked their lives and knew it.

Most telling of all, the men lacked any experience whatsoever in running the rapids that Pá-ri-ats and Powell's previous observations said lay ahead. The crew included mountain men, Civil War veterans, a printer, a boy of seventeen, and an adventure-seeking Englishman who just happened by, perhaps having arrived at one end of Green River as end-of-track departed the other. Powell had floated the length of the Mississippi, the Ohio River from Pittsburgh to St. Louis, and the length of the Illinois River. During the Mississippi trip, Stegner muses, he might have passed Sam Clemens, better known as Mark Twain, going the other way. But canoeing down those streams was as different from floating the Green and Colorado as a ride on a horse and buggy was from a trip on the locomotives then chuffing their way through Green River. On those placid eastern rivers, one knows what lies around the next bend. And should trouble occur, the sailor can pull into the bank and safety almost at will. But from the Uintas on, the Green and Colorado offer few safe havens and even fewer exits. Between Green River, Wyoming, and the mouth of the Uinta, 1,000 miles downstream, explorers had reached and forded the Green and Colorado at only three points: where the Uinta River enters below the mountains of the same name; at present-day Green River, Utah, where Captain John Williams Gunnison crossed in 1853 seeking an alternate route for the transcontinental railroad; and above present-day Lee's Ferry, Arizona, where Fathers Dominquez and Escalante forded in 1776 returning to Santa Fe, having failed to find a route to Monterey, California. Powell's crew was in approximately the same position as the greenest dude about to alight on the back of a bucking bronco: getting on would prove relatively easy; staying on and getting off at your volition, rather than that of the horse or the river, would prove difficult.

⚑

YET THE JOURNALS OF THE EXPEDITION members convey no fear— only exhilaration and an impatience to get underway. As the party sailed off down the Green River on May 24, 1869, around the first bend and out of sight, it might as well have floated off the face of the earth. No one heard a word until June 30, when the Corinne, Utah, *Reporter*, a railroad camp rag, reported that the entire party had perished in one of the awful rapids of the Green. The Omaha papers and then the *Chicago Tribune* picked up the story of the tragedy, the latter describing the extensive interviews given by the lone survivor, one John A. Risdon. The genuine tears in his eyes as he recalled the last moments of his shipmates lent credence to his tale and brought him sympathy from almost everyone who heard or read of it. The *Tribune* reported:

> On the 7th or 8th day of May, the party reached the Colorado River at a point called Williamsburg, a small Indian settlement. At that time the party consisted of Major Powell, William C. Durley, Charles Durley, Andrew Knoxon, T. W. Smith, William S. Dolton, Charles Sherman, Wm. Scott, Perry Duncan, John Jones, Frederick Buckingham, David Sellers, Edward Spencer, Wm. Murray, Isaac Thomas, Thomas Heughs, a half-breed named Chick-a-wa-nee, the guide, and two men who lived at Fairberry, Ill., who acted as runners, and whose names he could not remember. There were also two teamsters named Fred Myers and Thomas Welch.

Risdon even earned an audience with the Governor of Illinois, Powell's home state, who sized him up as "an honest, plain, candid man [who] told his story in a straightforward manner." But not quite everyone believed the heartbroken, lucky survivor. Emma Powell, waiting out the trip at her home in Detroit, pointed out that no one named Risdon had been among the crew. Moreover, she had received a letter from her husband dated May 22, four days *after* their reported demise. The men had posted the letter at the Uinta Indian Station, to which some of them had hiked, forty miles up the Uinta River south of the Uinta Range. (See Figure 2, page 15.) Inconveniently for Risdon, they sent their letters beyond the point and after the time at which he said they had perished. The comment that Sam Clemens later made would have been apt: "the report of my death was an exaggeration." Risdon's lie was so audacious that it inspired not only anger, but the kind of begrudging admiration reserved for a master fabricator like Baron Münchausen.

What is especially telling about Risdon's prevarication is that he invented not only a wholly mythical crew, but a wholly new geography for the Colorado Plateau. The death of the men came, he said, after they elected to explore two tributaries of the Colorado, the Delaban and the Big Black. To get to these side streams the party had to cross the main stem, which for some unexplained reason they chose to do not in their wooden boats, but in a large birch-bark canoe. All the men, that is, but Risdon, who had been sent to scout the Delaban from shore. He tried his best to dissuade his fellows, for here the river dropped 160 feet in a mile-and-a-quarter, nearly the height of Niagara Falls. But Major Powell had scoffed at the cautious Risdon: "We have crossed worse rapids than these, boys. You must be getting cowardly. If seven or eight men cannot paddle us across there, we will have to go under." And go under they did, "Captain" Powell at the helm of his vessel, in the best naval tradition going down with his ship. "For two hours," Risdon reported, "I lay on the bank of the river crying like a baby." How strange to find a crocodile shedding tears in Colorado! Indeed, not only did none of the named crew except the Major exist; the Delaban River and the Big Black were figments of Risdon's imagination. What better proof of how little known was the Plateau country than the immediate acceptance of the two rivers, no more real than Styx, by no less a distinguished newspaper than the *Chicago Tribune*?

Risdon disappeared from the pages of history as soon as he arrived. Vastly more significant is the failure of Powell and his men to drown on their way to the mythical Delaban. They made it to the Uinta Station, forty miles up the Uinta River, (see Figure 2) in time to mail their letters (including the one to Emma) and prove Risdon a liar. Indeed, all but three made it safely to their takeout at the mouth of the Virgin. Those three died, not drowned in the rapids, but on the plateau. They lost faith and, deep in the Grand Canyon and only two days from the end, at what we call Separation Canyon, elected to walk out. Somehow the three men reached the plateau thousands of feet above, but there a band of Shivwits Indians, mistaking them for the white men who had molested and shot an Indian woman, killed them.

Modern-day river runners have everything that Powell and his crew lacked. Today they often ride in all-forgiving rubber rafts rather than wooden boats, know in advance every twist, turn, and rapids of the river, and, in case they need assistance, are not the only party on the water. Today's river runners can only shake their heads in wonder that Powell's men would even try to run the river, and, even more unlikely, that they would succeed. That the voyage also made geological history, in spite of having only one person who could claim any expertise, is a tribute both to the acuity of their leader and to the unique terrain through which they traveled. Seldom has any party with such poor credentials accomplished so much. An inescapable conclusion is that they were ably led.

MAJOR JOHN WESLEY POWELL was a veteran of Grant's Army, an artillery officer who lost his right arm in the Shiloh Hornet's Nest. A footloose, vagabond scholar, he had learned more tramping the fields and floating the rivers of the Midwest than in any school. His interest in nature came as a boy; after he entered Oberlin College in 1857 he became an enthusiastic botanist, organizing his class into a club to assist in specimen collecting. Together they searched the woods and swamps to create an almost complete herbarium of the flora of Lorain County, Ohio. Powell's later nemesis, Ferdinand Vandeveer Hayden, had graduated from Oberlin in 1850 and thus did not quite overlap with the future Major. While a student, Hayden too became enthusiastic about botany. Perhaps at austere Oberlin, its bedrock buried by glacial drift and thus offering neither natural nor academic laboratories of geology, botany provided the only way to pursue an interest in natural history. In 1858, Powell returned to Wheaton, Illinois, where his family had settled, joined the Illinois State Natural History Society, and conducted a natural history survey of the state. In 1867, the Illinois Legislature appointed him Professor of Geology and Natural History at the State University at Normal. At about the same time, the Illinois State Natural History Society chose him as its curator. Powell's most irascible future companion on the river, self-proclaimed mountain man Jack Sumner, later said the idea of exploring the Colorado had been his. But like most of Sumner's statements, this one seems exaggerated. After all, it was

Powell who, in December 1867, told the Illinois State Board of Education that having seen the headwaters of the Colorado, he planned to "complete the exploration" of the river the following year.

Powell and a group of friends and family members spent the summers of 1867 and 1868 in Colorado, scouting the terrain and gathering specimens to augment the collections of the museum. He, his wife Emma, and several companions stayed at "Powell Bottoms" on the White River in Western Colorado during the winter of 1868. He explored the surrounding country, including the Green River area, and visited frequently with the local Ute tribe. As soon as he returned to Illinois that spring, he set his plan in motion.

Powell supplied his own answer as to the purpose of the expedition: "The exploration was not made for adventure, but purely for scientific purposes, geographic and geologic, and I had no intention of writing an account of it, but only of recording the scientific results." This was likely disingenuous. The Major surely knew that the successful navigation of the fearful canyons of the last unexplored region of the country would bring the leader national attention. But just as surely, never has there been any hint that Powell hoped to use that attention, or anything else that ever came to him, for financial gain. Given what he was to accomplish in Washington, it seems likely that he had unusual ambitions of public service. Whatever his motives, he became a "national celebrity," a "glittering public hero," and "the most celebrated adventurer since Lewis and Clark."

The motives of his men were a mix as well. They knew that the Green and Colorado drain mountain ranges where miners had found gold; the California fields had demonstrated how rivers concentrate pay dirt downstream in placer deposits. Who was to say that the unexplored Colorado had not done the same? Indeed, far downstream it had, but the gold proved so fine that Powell's crew could never have detected it. As was the case with Powell himself, when the voyage began, the diaries of the crew mention neither gold nor any other reward. The striking, dangerous country; their need to focus on the task at hand in order to survive; and dreams of sumptuous meals fully concentrated their minds. What these men mainly sought, especially the war veterans bored by peacetime, was adventure. They were the polar opposite of the Conquistadors, whose sole object was gold.

Powell's popular writings about the journey not only made him a national figure, they helped propel him at age forty-five to the directorship of the new Bureau of Ethnology and at forty-seven to succeed Clarence King as the second Director of the United States Geological Survey. The itinerant country boy, the self-taught Wes Powell, rose to become the equal of the most able men in the country. Few who never held elected office have wound up with more influence in Washington. But a single false move by a man suspended by his only arm, high on a canyon wall, and United States history would have been different.

South from Green River, Wyoming, the country consists of rolling, semi-desert plains. In his later writings, Powell would mark as the true beginning of the American West the 100th meridian, which lies about 250 miles west of Omaha. Westward from there, annual rainfall drops below the twenty inches needed for non-irrigated agriculture. At Green River, Wyoming, annual rainfall averages only about twelve inches and the lack of vegetation shows it.

Traveling south from the town, one sees looming in the distance, perpendicular to the southerly course of the Green River, a dark mass of mountains. The east-west orientation of the Uinta Range puzzles geologists, as almost all other mountain ranges in the Americas—the Appalachians, the Rockies, the Sierras, the Andes—trend in a north-south direction. Almost exactly 100 years after Powell, geologists were to discover what governs the configuration of the continents and the disposition of their mountain ranges: plate tectonics. The slow churning of the earth's mantle causes two plates at the surface to collide, throwing up a mountain range; elsewhere, a continent may rift apart to create two continents with a new ocean basin in between. The Appalachians, for example, mark the site where an ocean ancestral to the present Atlantic Ocean first closed, suturing together the continents on either side and deforming the rocks. Later, the mega-continent split apart and the two halves drifted away to leave mountain ranges on each separated by the Atlantic. The Appalachians run north-south because that is the direction of the original collision and the subsequent split—now marked by the mid-Atlantic ridge-rift system. Deciphering the history of the Rocky Mountains and the western edge of North America has proven more difficult, as have the precise reasons for the east-west direction of the Uintas. But all this lay far ahead of Powell and his small flotilla; before the theory of continental drift, the predecessor of plate tectonics, was even proposed, half a century would pass; another half a century would elapse before geologists accepted the theory.

At this stage in his incipient career, the romantic and realistic sides of Powell's nature tended to alternate. One moment his account waxes poetic over the scenery; the next it soberly assesses the crew's chances of making it through the rapids ahead and, if they do, of subsequently starving. In Powell's later years, an idealistic aspect of his character emerged. In the 1870s and 1880s, some boosters viewed every difficulty the West presented to settlement, including the low rainfall, through the rosiest of glasses. Some even claimed that "rain follows the plow": culti-vate the land and more rain will fall than if farmers had not broken the soil. Powell was to debunk this and other myths about the West; eventu-ally he alienated powerful western politicians whose support he needed but who found his pronouncements about the lack of water in the West in conflict with their own plans and ambitions.

Though rain does not follow the plow, it did plague the men in their first few days out of Green River town. But as getting soaked would soon be their daily lot, getting used to it did them no harm. In the sixty-mile stretch leading to the Uintas, the Green River runs swiftly but without white water to teach the novice how to run rapids. The looming moun-tains brought out the romantic side of Powell, who saw "high peaks thrust into the sky, and snow fields glittering like lakes of molten silver, and pine forests in somber green, and rosy clouds playing around the borders of huge, black masses; and heights and clouds and mountains and snow fields are blended into one grand view." No wonder the scene inspired him. Powell's native Midwest held nothing like this. Back home, moun-tains were absent and even bedrock was hard to find, obscured by soil, vegetation, and glacial drift. In the canyons he was about to enter, Powell would see nothing but raw rock—from now on it would be soil and vege-tation that were scarce. But raw rock is just what a geologist seeks.

In his explorations during the two years leading up to the expedition, Powell had visited the Green in several places near the Uintas. Where he could not reach the water, he had looked down upon it from cliffs high above. Thus, he well knew that the Uintas lay directly athwart the course of the river. But rafting south toward the mountains, he found that the river glided "on in a quiet way as if it thought a mountain range no formidable obstruction." Then the Green

Enters the range by a flaring, brilliant, red gorge, that may be seen from the north a score of miles away. The great mass of the mountain-ridge through which the gorge is cut is composed of bright vermilion rocks; but they are surmounted by broad bands of mottled buff and gray, and these bands come down with a gentle curve to the water's edge on the nearer slope of the mountain. This is the head of the first cañon we are about to explore. We name it Flaming Gorge. The cliffs or walls we find to be about one thousand two hundred feet high.

Powell knew that at Flaming Gorge, the river enters a "canyon cut nearly halfway through the range, then turns to the east and is cut along the central line, or axis, gradually crossing to the south. Keeping this direction for more than 50 miles (through Brown's Hole or Park, see Figure 2, page 15), it then turns abruptly to a southwest course, and goes diagonally through the southern slope of the range [through Lodore Canyon]." The initial incursion of the river into the mountains provided the first real white water the men had ever breasted: "I stand up on the deck of my boat to seek a way among the wave-beaten rocks. All untried as we are with such waters, the moments are filled with anxiety. Soon our boats reach the swift current; a stroke or two, now on this side, now on that, and we thread the narrow passage with exhilarating velocity, mounting the high waves, whose foaming crests dash over us, and plunging into the troughs, until at last we reach the quiet water below. Then comes a feeling of great relief. Our first rapids is run." Before their trip ended, they would run hundreds, many of them more fearsome than the infant rapids they had just survived. Exhilaration would pass into boredom, and then into fear.

No sooner had the Green begun to penetrate the Uinta Range beyond Flaming Gorge than it "wheels back on itself, and runs out into the valley from which it started . . . the canyon is in the form of an elongated letter U, with the apex in the center of the mountain. We name it Horseshoe Canyon." Soon the cliffs towered 2,500 feet above them, and the water of Red Canyon, as Powell named this stretch, roared continuously. A few days later they entered Brown's Park, a hidden paradise known to trappers like Jim Bridger, to the ranchers of Powell's day, and later, to outlaws Butch Cassidy and the Sundance Kid, as well as drovers on their way from Texas to the Wyoming railhead. It made a fine spot for the crew to rest and prepare themselves for the plunge that the Green was about to take into the heart of the Uintas. From here on, the terra would be truly incognita.

As he did whenever he could, Powell, together with some of the crew, climbed atop the surrounding cliffs—now to the very spot he had reached the previous spring when reconnoitering with Emma. There he sat, his legs dangling over a 2,000-foot precipice, a feat that he could accomplish only after several years of practice had "cooled his nerves." By the time he came down at noon, he found that "the sun shines in splendor on vermilion walls, shaded into green and gray; the river fills the channel from wall to wall, and the canyon opens, like a beautiful portal, to a region of glory." But later in the day, as the sun set, these pleasing hues slowly changed. Now they were "somber brown above, and black shadows are creeping over them below; and now it is a dark portal to a region of gloom—the gateway through which we are to enter. What shall we find?" One rafter said that the canyon that lay just down river resembled "a mountain drinking a river." An unromantic member of the crew wrote that this new canyon "looks like a rough one for the walls are very high and straight and the sides are of sand-stone much broken with seams but at the mouth nearly perpendicular; in such the worst bowlders have been found and I expect them below here."

The youngest member of the party, recalling a poem learned at school about a rushing stream, named it the Canyon of Lodore. A modern river runner wrote, "Each canyon of the Green has its own distinctive presence but none is as dramatic. The cliffs rise two thousand feet, immediate, all the more striking because of the pale landscape from which they spring, almost without transition. The Gate of Lodore hinges inward, cruelly joined, hard rock, ominous, and when mists skulk low between the cliffs, they become an engraving by Gustav Doré for one of Dante's lower levels of hell." The comparison with Hades proved apt: before the men got out of Lodore they had capsized and lost one of their four boats and most of its cargo, amounting to one-third of their total supplies. It appeared at first that they had lost their precious barometers as well. Although Powell had required each of the three instruments to be on a different boat, somehow all had wound up aboard the sunken *No Name*. The barometers were critical not only to science but to their peace of mind. They had no map. They could try to fix their geographic position—their longitude and latitude—with a sextant, and from that estimate how many miles they had to travel if they had been crows. But they floated down a river that they could already see followed a tortuous path. The barometers added a vital kind of information, allowing them to measure their present elevation and compare it with the known elevation of their destination, the mouth of the Virgin River. Thus, they could find out how far they had to descend. In the Canyon of Lodore, the method told them that they had thousands of feet to drop, and, they fervently hoped, many miles laterally in which to accomplish it. What did not need to be stated again was their earnest wish for a gradual, steady decline, rather than one in which the river accomplished scores of feet of its fall at once.

As good fortune would have it, the men were able to recover not only the capsized barometers, but miraculously, a full jug of whiskey, which some unknown party had smuggled aboard. Before long, the cook had set the willows afire, requiring the men to abandon much of their kitchen gear and spring for their lives into the remaining boats and escape downstream. Powell the optimist joked "water catch 'em; h-e-a-p catch 'em; we do just as well as ever." Looking back, Powell the realist called the Canyon of Lodore "a chapter of disasters and toils." They had packed food for ten months, lost a third of it almost as soon as they hit white water, and before their journey ended three months later would find their diet reduced to coffee, flour, and mangy dried apples.

The modern rafter, launching just above the Gate of Lodore, joined by whole families of rafters and children piloting their own kayaks, is apt to find the previous ominous description overdone. With Flaming Gorge Dam having tamed the Green, and the requirements of a world-class fishery for a steady flow below the dam having largely eliminated the possibility of high water, the chance of disaster is remote. One can as easily find the Gate of Lodore downright inviting—a sort of geological pearly gates. Certainly the modern geologist can hardly wait to pass through and enter the hallowed ground of geology, where one can still see what Powell saw and ponder the great geological principles that first began to take shape in his mind here.

For one attempting to understand how Powell came to discover those principles, a question that immediately arises is how much geology he knew at the start of the trip versus how much the raw country taught him its principles as he rafted and observed. His intelligence and ability to see further than most are not in doubt; otherwise, this largely self-educated country boy could not possibly have risen to become one of the most influential scientists of his day and the most influential non-elected man in Washington, D.C. But how much did he know of geological debates at the time when he began his voyage? In his teaching, Powell had used the texts of Dana, Agassiz, and others, though like any geology professor, he preferred to take his students outdoors. During the long days of the Civil War he had read all the geology he could get his hands on. There seems no doubt that he did know of the geological controversies of his day, in which the question of which came first, the river or its valley, figured large.

Now the observant, well-read young professor found himself in a place where erosion has exposed naked rock for hundreds of miles in two dimensions and thousands of feet in the vital third. European geologists debating the origin of river valleys and landscapes had far more training and loftier reputations. With little of either, John Wesley Powell had the advantage on them. He used it not only to confirm that rivers carve their valleys, not only to invent a theory explaining how the Green and Colorado rivers had incised their canyons, but to discover broad principles that govern rivers everywhere. To be able to understand how he did so, we need to review the state of the science of geology by the mid-nineteenth century and how it had gotten to that state.

Chapter 3

How Old Is a River?

T o one with Powell's curiosity about the world, rivers were irresistible. Indeed, natural historians have always paid attention to rivers, and for good reason. Not only are rivers important in their own right, they entwine so intimately with landscapes that, as author Patrick McCully recommends, we ought to refer to "riverscapes." Though it took geologists a while, eventually they came to recognize that stream erosion has been the dominant instrument in shaping the surface of the earth, far more so than wind and ice erosion. Even in an area like the American southwest, where it seldom rains, water remains the overriding geologic agent.

Rivers hold only about one-four-thousandths of the earth's surface water. Yet at the average rate at which they lower the land surface, in only twenty-five million years, rivers could have reduced every continent to sea level. Because the earth is almost two hundred times older than that, why do any continents still exist? In the answer hangs a fundamental truth about the earth, one that we will come back to later.

For the genus *Homo*, rivers have been a matter of life and death. The oldest hominid fossils come from the banks of the Awash River in Ethiopia. The first archeological evidence of the shift from hunting and gathering societies to fixed settlements occurs at sites nine to ten thousand years old along river valleys. Other ancient sites adjoin the Indus,

Ganges, and Huang Ho (Yellow) rivers. The great ancient cities of Ur, Ashur, Nimrud, Nineveh, Samarra, and Babylon all developed along the rivers of Mesopotamia ("between the rivers"), as others did later beside the Nile and elsewhere.

The reasons are obvious. Rivers provide water for drinking, washing, and irrigation; plants and animals for eating; and currents to carry waste downstream. Trees and bushes along riverbanks supply shelter and fodder, especially during drought. Throughout history, rivers have offered the easiest routes for explorers, traders, settlers, and conquerors. When Lewis and Clark sought a route to the Pacific Ocean, they went not by land but as far as they could up the Missouri, all the way to its headwaters, and, when over the spine of the Rockies, descended the Columbia to the western sea. Well into the era of the railroad, rivers still provided the principal means of transportation. One can only wonder how far, without rivers to integrate and sustain them, human societies would have progressed beyond a nomadic state.

GIVEN THE IMPORTANCE OF RIVERS and their proximity to centers of culture and learning, and especially because of their possible connection to the Biblical Flood, it was only natural that early geologists would ponder how rivers and their valleys came to be. The most fundamental question was whether rivers erode their own valleys, or whether some process pre-constructs the valleys, into which the rivers then flow. Which came first, the river or the valley?

But before one can even ask such a question, one has to be able to conceive that rivers, valleys, mountains, plains, and all the features of the earth—even the earth itself—have a history. The realization that they do goes back only a few hundred years; for all of human history before that, if people gave the matter any thought, most assumed, according to their own creation myth, that the earth and all its natural features had been created instantaneously and for them.

Scholars from Greek, Arab, Chinese, and Egyptian societies had begun to use observation and reason to comprehend that instead, the earth has a history that can be understood. But in one of the great tragedies of human history, much of what the ancients deduced, the rise of Christianity submerged, requiring scholars to make the same discoveries again centuries or millennia later. For instance, in the fifteenth century, Leonardo da Vinci's contemporaries believed that fossils were tricks played by God to test our faith, so-called "sports of nature." He, in contrast, recognized fossils as the remains of once-living creatures. But 2,000 years before Leonardo, Xenophanes of Colophon had noticed that shells found in rocks from the mountains of Attica closely resembled those of creatures then living in the Aegean Sea. He concluded that the shells were the preserved remains of clams from an earlier period when Attica was under the sea. Xenophanes clearly understood that the earth has a history. Herodotus used the right method to get the wrong answer when he decided that the tiny objects embedded in Egyptian limestones, which we recognize as the shells of fossil foraminifera, were the petrified remains of lentils tossed out by the pyramid builders after an alfresco lunch. But for a millennium, all such reasoning vanished.

Scientific progress rests on an accumulation of observations, facts, and theories. This makes it difficult to read the history of a field and point to a single advance as the most important of all. But in geology, it may be possible to do just that. A science of the earth depends on understanding that the earth has a history. Had our world been created instantaneously, there would be no way to even conceive of geology. A Danish anatomist named Niels Stensen (1638–1686), his name Latinized to Steno, was the first to recognize that our planet does have a history and that by using careful observation and logic, we can discover that history. Steno was not only the first geologist, but the first paleontologist as well.

In the mid-seventeenth century, 150 years after the death of Leonardo da Vinci, Steno came under the patronage of Grand Duke Ferdinand of Tuscany. In October 1666, two fishermen, on Ferdinand's orders, brought Steno the head of a giant shark that they had caught near Livorno. Steno made meticulous observations of the shark's teeth and concluded that they were identical to the delta-shaped, mysterious "tonguestones" found embedded in certain rocks. While one of his contemporaries had attributed fossils to a "lapidifying virtue diffused through the whole body of the geocosm," Steno concluded that the tonguestones looked so much like shark's teeth that they must have once been shark's teeth, which had since become petrified. He wrote, "Nothing seems to contradict the theory that the bodies excavated from the earth and which resemble parts of animals must also have been parts of animals." This might be said to be the founding concept of paleontology.

For a rock to encase a fossil tooth, which Steno classified as "a solid within a solid," the tooth must have existed before the rock began to form around it—the tooth must be older than the rock. Today this would be a trivial observation, but in Steno's day it was a revelation, for it showed that some natural objects were older than others, which can only mean that all were not formed in the same instant. Where others saw sports of nature and the effects of a lapidifying virtue, Steno saw a vanished ocean, teeming with sharks and their prey. He saw an earth with a history—with time—albeit the short, Biblical time scale of his day. The recognition of time in earth history might be said to be the founding of geology. Ever since, geology and paleontology have had a symbiotic, reinforcing relationship, research in one leading to insights in the other.

Steno's scrutiny of the tonguestones led him to study not only fossils, but other natural solids such as minerals, crystals, incrustations, veins—even entire rock layers. From the last he deduced three fundamental principles on which geology has depended ever since. Steno recognized that sediments settle onto the bottom of a body of water and therefore accumulate in horizontal layers. Where we find beds that are not horizontal, some force must have tilted them. That sediments settle from a fluid means that in a sequence of layered rocks, as in a brick wall, the bottom layer is the oldest. This is the Law of Superposition. The lower a rock layer, the older it is, again showing that rocks record a time sequence and that the earth has a history. Steno's third principle was that horizontal beds extend laterally in all directions. The same beds found exposed on either side of a valley must have once been continuous. We will meet this idea again shortly.

Using his principles, Steno worked out the geologic history of Tuscany, in the process making the first geologic cross sections. Those who came after Steno used his Law of Superposition to decipher the geologic history of other areas, to correlate rocks across distances, and to make maps that showed different rock formations. By understanding fossils and the meaning of layered rocks, and by recognizing that the earth has a history, Steno captured the essential principles that underpin the science of geology. Two centuries later, Darwin would use geology and paleontology to buttress his theory of evolution. Remarkably, Steno did all his scientific work in only a few years. He then converted to Roman Catholicism and abandoned science.

EARLY GERMAN GEOLOGISTS studying the Alps and Urals discovered that the lowest, oldest rocks of a sequence tended to be hard, crystalline, and unlayered. Above these rocks lay beds that contained fossils, and atop those were unconsolidated sand, silt, and gravel. The geologists found the same sequence in other, widely separated areas (and we find it in the Grand Canyon as well). Were these everywhere the same set of rocks? If so, what set of events in earth history did they reveal? Why were the oldest rocks hard and crystalline and the youngest loose and sandy? Just as the stage was set to answer these questions by observation and deduction, many decided that the Bible had already provided the answer.

By the 1600s and 1700s, whether or not a scientist was himself devout, and most were, any theory of earth history had to conform to the Biblical account of creation and, in particular, to the great Flood of Noah. By 1701, an authorized Bible contained the chronology of earth history worked out by James Ussher (1581–1656), Archbishop of Armargh and later Primate of All Ireland. Ussher was one of the most learned men of his day, having been a professor and Chancellor at Trinity University in Dublin. He used the Bible, as well as his substantial library, to work backwards and calculate that God created the earth in 4004 B.C. People soon came to regard this date with the reverence usually reserved for the Bible itself, as though God, and not Ussher, had specified the date. The King James Bible introduced into evidence in the Scopes "Monkey" trial in 1925 contained Ussher's chronology.

The few thousand years that Ussher's timetable allowed for all of earth history obviously has great implications. If the earth is so young, all the features that we see on its surface—mountain ranges, rivers, plains, deserts, and so on—must have arisen almost instantaneously. Earth history then was one catastrophe after another, rendering Steno's insights trivial. The Bible did not spell out that history. But it did tell of one specific catastrophe with geological implications: the Flood.

THE MOST PROMINENT OF THE EARLY geologists to square field observations with the Bible was Abraham Gotlob Werner (1750–1817), a professor at the School of Mining in Freiberg, Germany. An expert mineralogist and brilliant teacher, Werner inspired such devotion in his students that, even after others had falsified his theories, his former students stood by him. Werner was the archetype of the "armchair geologist" who does no fieldwork but relies on that done by others. Indeed, he traveled hardly at all, satisfied that what he saw around him in Saxony had universal application.

Werner reconciled the newfound geologic observations and the Flood by proposing that all rocks, no matter what type, precipitated from a universal ocean, in the tripartite order that others had found in the Alps and elsewhere. Even the crystalline rocks at the bottom of the typical sequence, Werner said, had never even been hot, much less molten. So-called lava was not volcanic, but coal beds that had caught fire and burned. As the waters of the great ocean receded, they sculpted the land into the shapes and features that we find today. Werner's theory ignored the interior of the earth and did not say where the disappearing waters went, still a problem for those who espouse "flood geology."

Werner not only stayed home, he stopped reading his mail. The distinguished Académie des Sciences sent a letter telling Werner they had elected him a member, but he never got around to opening the letter and learned of his honor only years later. He also preferred not to write and authored only some twenty short papers. What Werner evidently did like to do was teach, and it is his prowess as a professor and the loyalty he inspired, rather than his failed theory, for which we ought to remember him.

By the end of the eighteenth century, advances made during the Renaissance, the Protestant Revolution, the Enlightenment, and the Industrial Revolution brought about a corresponding revolution in human thought and in the approach to the natural world. It was only a matter of time before, in interpreting the earth, scholars began to rely more on observation and reason than religious dogma.

SCOTSMAN JAMES HUTTON (1726–1797) developed the theory that opposed and eventually defeated Werner's ideas. Hutton trained as a lawyer, a chemist, and a physician but became a farmer, which allowed him to remain close to the earth and his passion, geology. His intellectual circle in Edinburgh included Joseph Black, a founder of modern chemistry; James Watt, inventor of the steam engine and at one time Hutton's laboratory assistant; Adam Smith, author of *The Wealth of Nations*, who served as Hutton's executor; and mathematician John Playfair, who became his interpreter. John F. Kennedy once remarked at a White House gathering, "I think this is the most extraordinary collection of talent, of human knowledge, that has ever been gathered together at the White House, with the possible exception of when Thomas Jefferson dined alone." Among the men gathered at Hutton's table, Jefferson would have been right at home.

Hutton's methods were to make one rock outcrop the most important of all. At Siccar Point, in Berwickshire, Scotland, a set of nearly horizontal sedimentary layers lies above another set whose beds are vertical.

Where the two meet, the surface is rough and irregular. To demonstrate this idea, hold one palm toward you with your fingers (representing the beds) horizontal, and then place the fingers of your other hand vertically beneath it. Steno's logic had shown that sedimentary rocks accumulate in horizontal beds; therefore, when we find them tilted on edge, we know that some sort of upheaval lifted and rotated them.

But Hutton was able to read much more into the scene at Siccar Point. Both sets of rocks are sedimentary, meaning that they comprise fragments eroded from older rocks and deposited under water. Hutton reasoned that the older, bottom sequence formed when such fragments collected in horizontal layers; the weight of overlying sediments then compressed them into solid rock. Next, some geologic force stood the layers on end and lifted them above the sea. There they eroded and eventually sank back beneath the waters again. At some indeterminate time later, new sediments eroded and accumulated in horizontal layers on top of the older rocks. These new layers also hardened into rock, and then both sequences again rose above the sea. The rocks exposed at Siccar Point today are eroding, producing fragments that will wind up in some future rock.

Hutton drew from this scene two conclusions of utmost importance. In a single human lifetime, Siccar Point changes not at all. Therefore, to encompass everything that he deduced had happened there, he knew that the earth must be much older than a few thousand years. As to how old, no one could say. Second, rocks erode to generate fragments that wind up in other rocks, which themselves are eroded to provide sediment for the next generation, and so on and so on, in a great rock cycle. A given sedimentary rock gives no clue as to how many such cycles it has passed through. Hutton could only say that he could find "no vestige of a beginning; no prospect of an end."

With no hint of its commencement, geologic time was to Hutton an "abyss"—one deep enough to accommodate all of earth history. No longer was it necessary to compress all that had happened into the few thousand years of Ussher's Biblical chronology. The problem Hutton's time scale produced for fundamentalists, and still does, is that an earth so old must have existed for a long time before man appeared, making it less probable that God created the world specifically for us.

Thus freed from the restraints of a young earth, Hutton saw that the common processes that we see going on every day—erosion, transportation, deposition, glaciation—could explain geological observations. Not only was Hutton able to appreciate the true grandeur and meaning of a scene like Siccar Point, he drew from it a universal principle: to understand what happened in the past, one need only observe the processes going on today: The Present is the Key to the Past. Like most slogans, this one is oversimplified and, carried to an extreme, does more harm than good. Later it became proscriptive, interpreted to mean that nothing that we do not see going on today could have taken place in the past. Giant ice sheets could not advance thousands of miles and cover continents; meteorites could not fall and affect the earth; continents could not drift. Nevertheless, Hutton's theory of uniformitarianism, as others came to call it, was, after Steno's insights, the fundamental intellectual breakthrough in the history of geology.

Werner's theory of a universal ocean survived for a few decades after Hutton, kept alive by allegiance to God and to Werner. But though geologists eventually gave up on Werner's idea of precipitation from a universal ocean, most retained the notion of a short time scale filled with cataclysms. These catastrophists were emboldened by several geological observations that the Huttonians could not explain:

First, mountains often contain rocks that some geologic process has folded, broken, and metamorphosed beyond recognition. But no process that we see going on today causes such deformation.

Second, scattered across Britain and Northern Europe, and especially in the valleys and mountain slopes of Switzerland, perch boulders that can be as large as a small house. The rocks that make up these erratic stones often come from sources hundreds of miles away. Present-day streams lack the energy to move such giant boulders.

Third, streams often flow in valleys much larger than the streams themselves. Because, the catastrophists believed, the earth is only a few thousand years old, such small streams have not had time to carve their large valleys.

Hutton's theory had no answer for these questions, but, many believed, the Bible did. Catastrophes could create mountains; floodwaters could move boulders long distances, strand them atop Alpine ridges, and carve deep valleys into which subsequent streams would naturally flow.

GEORGES CUVIER (1769–1832), was born in Switzerland but adopted France as his homeland. He used Hutton's methods to reach opposite conclusions. Cuvier was an anatomist whose meticulous study of the rocks of the Paris Basin proved that some animals—giant sloths, enormous salamanders, and bizarre elephants—no longer exist. Some believed that explorers would yet find living representatives of such creatures. Indeed Jefferson, and Lewis and Clark, thought that Mammoths might still roam the unexplored American west. But Cuvier said they had disappeared forever—they had become extinct. He and his colleagues found strata that repeatedly alternated between marine and freshwater conditions, apparently supporting Hutton's notion of repeating cycles of earth history. But Cuvier parted company with Hutton by concluding that a series of "revolutions" have marked our earth. As the land repeatedly rises above the sea and falls beneath it, one set of organisms is destroyed while another emerges to take its place. Once a species has become extinct, it can never reappear—extinction is forever, as the modern saying goes. Though he was a lifelong Protestant, Cuvier did not identify his revolutions with specific events in the Bible. But others did, associating the last revolution with the Flood.

Though Cuvier's idea of successive revolutions, or catastrophes, gave way to Hutton's uniformitarianism, more recently the evidence of asteroid and comet impact as a geologic force has brought catastrophism back into repute. Geologists will likely always debate the relative importance of uniformity versus catastrophism, but that need not detract from Cuvier's major contribution to science: his discovery of extinction.

THE LORE OF GEOLOGY HAS IT THAT Hutton wrote so murkily that his more lucid friend, John Playfair (1748–1819), had to explain what he meant. Indeed, the scientific style of two centuries ago does make us itch to get out our editorial pencil. Yet when we actually read Hutton, though his prose is more ornate and circumlocutious than ours, his point usually comes across. Geologist and author Anthony Hallam quotes the following passage as evidence. The prose may appear antiquated, but is Hutton's point not clear?

> In examining things present, we have data from which to reason with regard to what has been; and, from what has actually been, we have data for concluding with regard to that which is to happen here after. Therefore, upon the supposition that the operations of nature are equable and steady, we find, in natural appearances, a means of concluding that a certain portion of time to have necessarily lapsed, in the production of those events of which we see the effects.

Nevertheless, Playfair was one of the clearest writers of any age. Here is what he had to say about rivers and their valleys:

> It is, however, where rivers issue through narrow defiles among mountains, that the identity of the strata on both sides is most easily recognized, and remarked at the same time with the greatest wonder . . . there is no man, however little addicted to geological speculations, who does not immediately acknowledge, that the mountain was once continued quite across the space in which the river now follows; and if he ventures to reason concerning the cause of so wonderful a change, he ascribes it to some great convulsion of nature, which has torn the mountain asunder, and

opened a passage for the waters. It is only the philosopher, who has deeply meditated on the effects which action long continued is able to produce, and on the simplicity of the means which nature employs in all her operations, who sees in this nothing but the gradual work of a stream, that once flowed as high as the top of the ridge which it now so deeply intersects, and has cut its course through the rock, in the same way, and almost with the same instrument, by which the lapidary divides a block of marble or granite.

Hutton and Playfair reasoned from observation rather than from dogma and backed up their conclusions with evidence:

Every river appears to consist of a main trunk, fed from a variety of branches each running in a valley proportional to its size, and all of them together forming a system of valleys, communicating with one another, and having such a nice adjustment of their declivities, that none of them join the principal valley, either on too high or too low a level, a circumstance which would be infinitely improbable if each of these valleys were not the work of the stream that flows in it.

If, indeed, a river consisted of a single stream without branches, running in a straight valley, it might be supposed that some great concussion, or some powerful torrent, had opened at once the channel by which its waters are conducted to the ocean; but, when the usual form of a river is considered, the trunk divided into many branches, which rise at a great distance from one another, and these again subdivided into an infinity of smaller ramifications, it becomes strongly impressed upon the mind that all these channels have been cut by the waters themselves; and that it is by the repeated touches of the same instrument that this curious assemblage of lines has been engraved so deeply on the surface of the globe.

It is there, says Dr. Hutton, that I would wish to carry my

reader, that he may be convinced, by his own observation, of this great fact, *that the rivers have, in general, hollowed out their valleys* [Playfair's italics].

Like the views of other prophets, those of Hutton and Playfair failed to find honor in the land of geology. Decades would go by before geologists accepted the wisdom of their view of earth history.

THE WORK OF REVEREND WILLIAM BUCKLAND (1784–1856), coming two decades after Playfair had elucidated Hutton, showed how little immediate effect the pair had. Like several of his contemporaries, Buckland was both a minister and a leading geologist and had reconciled the two callings. To accept Hutton would have required Buckland to renounce both his science and his faith. But then, what would he have left?

Buckland was one of the most popular lecturers at Oxford, no doubt in part because of his peculiarities. Indeed, on the all-time list of British eccentrics—and a long list it is—Buckland would rank near the top. Years after his death, his son told a friend of a visit the family had made to a famous foreign cathedral. Unlike Werner, Buckland was an inveterate traveler. The floor displayed stains that were perpetually fresh and that no one could remove; some said they were the blood of a martyr. But according to his son, "The professor dropped on the pavement and touched the stain with his tongue. 'I can tell you what it is; it is bat's urine!'" Buckland claimed to have eaten a member of every animal family, including not only mice baked in batter, but the embalmed heart of a French king.

In Buckland's 1820 inaugural lecture, entitled, "The Connection of Geology with Religion Explained," he set forth his beliefs clearly:

In [geology] we find such undeniable proofs of a nicely balanced adaption of means to ends, of wise foresight and benevolent intention and infinite power, that he must be blind indeed, who refuses to recognize in them proofs of the most exalted attributes of the Creator.

Like Playfair, Buckland focused on the valleys that are the most difficult to explain by catastrophism: those whose opposing bluffs show flat-lying, undisturbed sedimentary layers that match up across the valley, as in the Grand Canyon. Faulting, or some other catastrophic event, cannot explain these valleys, for they "bear no mark of having been moved from their original position by elevation, depression, or disturbance of any kind." Instead, something must have excavated the material that had once filled the space between the opposite bluffs. But where Playfair saw "nothing but the gradual work of a stream," Buckland saw removal of the rock that previously filled the valley by "violent and transient inundation." In other words, Buckland thought that a great flood had removed the rock. One paragraph in particular reveals how Buckland begged the question:

It is not easy to imagine how valleys of this last description [the kind described previously] could have been formed *in any conceivable duration of years* [italics added] by the rivers that now flow through them, since all the component streams . . . owe their existence to the prior existence of the valleys through which they flow.

Buckland could not conceive that enough years had passed for the slow geological work described by Hutton and Playfair, because to do so would violate his religious belief that the earth is young. Religion first, observation second. Valleys first, rivers second.

Buckland and his contemporaries found further evidence for their beliefs in the boulders scattered erratically across northern lands, even stranded halfway up the side of a hill, where present-day streams could not possibly have carried them. In Buckland's day, only one known geologic agent had the energy to move a boulder the size of a cottage: running water. Admittedly, present rivers were too small to have accomplished that task, but the Bible told of great waters that might have.

Several decades later, Alexander Agassiz identified a force which no geologist of Buckland's era could have imagined: huge ice sheets a half-mile or more thick that had crossed continents, carrying and, when they melted, dropping the giant boulders. In discovering the existence of past Ice Ages, Agassiz was one of the first to show how observation and reason allow cause to be deduced where we see only effect, the triumph of geology and of science generally.

⬛

AS THE NINETEENTH CENTURY progressed, the mantle of leading geologist passed from Buckland to his former student, Charles Lyell (1797–1875). Though he began as a lawyer, the success of his Principles of Geology, first published in 1830, allowed Lyell to devote the rest of his career to geology. He carried Hutton's ideas so far that the director of the Geological Survey of the United Kingdom, Sir Archibald Geikie, called him "the high priest of uniformitarianism," the name that had been given to Hutton's theory. Lyell's book appeared in twelve editions, the last in 1875, forty-five years after the first. Surely it is the most influential book in the history of geology. Lyell's legal training had taught him to frame an argument; his client was uniformitarianism, and for it he provided an elegant and persuasive brief.

By editing and re-editing his great book, Lyell could discard arguments that had proved wanting and incorporate new ones, all in such a way as to make it seem that he had known of them all along. Because few people other than historians took the trouble to read the earlier editions, Lyell appeared perpetually right, almost omniscient. But the approach carries the risk that instead of accepting a new idea for which the evidence is strong, one is tempted into papering over and retaining an obsolete old one. Even after Agassiz showed a doubting Lyell the overwhelming evidence for the Ice Ages, Lyell delayed accepting the theory as long as possible and then went out of his way to avoid giving Agassiz the credit.

Lyell's eminent position meant that he could not merely ignore new evidence, especially any that seemed to threaten his own ideas. Instead, he had to counterattack forcefully. In debates with two of the greatest scientists from other disciplines, he more than met his match. Outdoing Hutton, Lyell claimed that the earth is both infinitely old and a perpetual motion machine, both scientific impossibilities according to an even greater authority, British physicist Lord Kelvin. Though the earth turned out to be much older than Kelvin thought, its age is finite and the planet is subject to the laws of thermodynamics, the second one of which Kelvin discovered. Lyell's other long-running disagreement was with Charles Darwin, whose theory of evolution violated Lyell's notion of an unchanging earth. After years of dithering, in the last edition of his *Principles*, Lyell finally accepted evolution.

But in the next-to-last edition, published in 1872, a half century after Buckland had written on the subject and seventy years after Playfair had made the case, Lyell still had not accepted the argument that rivers carve their own valleys:

> It is probable that few great valleys have been excavated in any part of the world by rain and running water alone. During some part of their formation, subterranean movements have lent their aid in accelerating the process of erosion.

One reason that European geologists did not give credence to the ability of rivers to carve their valleys was that they had come to believe that marine erosion was far more important than fluvial erosion by streams. They believed the surface of the earth had mostly been shaped at a time when it was submerged beneath the sea. Ocean waves and currents had determined the topography, after which the land rose to expose the result.

To get a bit ahead of our chronology, but to illustrate the depth of feeling of some geologists, let us turn to Clarence King, a graduate of Yale's Sheffield Scientific School and the first Director of the U.S. Geological Survey. In 1877 he attacked uniformitarianism as the mark of "an army of scientific fashion followers who would gladly die rather than be caught wearing an obsolete mode or believing in any penultimate thing." He charged:

> Uniformitarians are fond of saying that give our present rivers time, plenty of time, and they can perform the feats of the past. It is mere nonsense in the case of the cañons of the Cordilleras. They could never have been carved by the pygmy rivers of this climate to the end of infinite time.

To find out just how long geologic time might be, King set up his own laboratory and calculated that the earth could not be older than 24 million years, a figure that Lord Kelvin endorsed as superior to his own estimate, yet comfortably close to it. The uniformitarians needed billions of years, but as late as the 1890s King and Kelvin would only grant them millions. Thus, for some, the evidence that rivers carve their valleys, founded on years of hard-won observation, paled beside the seemingly higher authority of Lord Kelvin.

In that same year in which Lyell's penultimate edition came out, 1872, John Wesley Powell, a still young man lacking formal education and standing in geology, of whom Lyell had surely never heard, completed the second of two voyages down an unknown river through the last unexplored region of the United States. Powell's courage, his intellect, and the raw country through which he traveled, quickly put paid to the outdated notions of Buckland and his successors. Geikie would soon write, "Had the birthplace of geology lain on the west side of the Rocky Mountains, this controversy [over whether running water, rather than ocean waves and currents, has shaped the land] would never even have arisen."

Powell's first task was to answer the question that he confronted as soon as he entered the Canyon of Lodore: "Why did the river run through the mountain?" After he dealt with that, he and the other pioneer western geologists could not only lay to rest the passé notion of valleys first, rivers second—they could discover the great principles that govern the evolution of rivers and riverscapes. And so let us begin to trace the efforts of the best American geologists to answer the most difficult geologic question on their continent. We will proceed as Powell did, drawing the lessons of the Green River before we move on to try to answer the great question of the cause of the Grand Canyon.

Chapter 4

The Saw That
Cut the Mountain

‹‹‹

Powell's rudimentary formal education had left him unencumbered with allegiance to any particular theory about how the earth works. Although he was a religious man, he never demonstrated a need to make earth history jibe with Genesis. Powell made geological observations without preconceived notions or the need to find evidence for some previously announced theory or dogma: he had written no scientific papers and espoused no theories. His mind was an open book on which the canyons of the Green and Colorado could write an indelible impression.

Powell had observed how the Green River wanders south over the high desert as it approached the northern edge of the Uinta Range (see Figure 2). At Flaming Gorge, the river cut briefly into the mountains, turned east, nearly looped a loop at Horseshoe Canyon (not shown in Figure 2), proceeded further east through Brown's Park, and then, at the Gate of Lodore, turned ninety degrees to the south-southwest and plunged straight through the mountains and out the other side. While the detailed reasons for this circuitous path took geologists a century to understand, in no way did the valley of the Green River resemble one cut by a great flood, into which the river subsequently just happened to find itself.

The drawing of the Gate of Lodore in Powell's popular account of his expedition illustrates a crucial difference between the rivers of the

Colorado Plateau and those of wetter climates like Powell's Midwest and Great Britain. Where rainfall is abundant, rivers tend to run in wide valleys floored with their own alluvium. Indeed, it was in part this disparity in size that led Buckland and others to conclude that, in the few thousand years they interpreted the Bible to say were available, streams could not have cut valleys so much larger than themselves. But in the Canyon of Lodore, not only is the valley not wider than the river; there is no valley. Call Lodore what you will—arroyo, canyon, chasm, cleft, defile, gorge, gulch, rift—a "valley" it is not.

In a typical river in a humid climate, bluffs on either side enclose the floodplain, comprising the sand, silt, and gravel—the alluvium— that the river laid down and redistributed as it meandered back and forth across the valley floor. If bedrock is visible anywhere, it is on the faces of the steep bluffs that bound the valley and rise to the countryside beyond. Playfair and Buckland had both used these exposures to bolster

Figure 3. *Powell's 1875 Cross-Section. Section from west to east across the plateaus north of the Grand Cañon, with bird's-eye view of terraces and plateaus above. Horizontal scale, 16 miles to the inch, vertical scale, 4 miles to the inch.*

Pine Valley
Mountain

Virgen
Valley

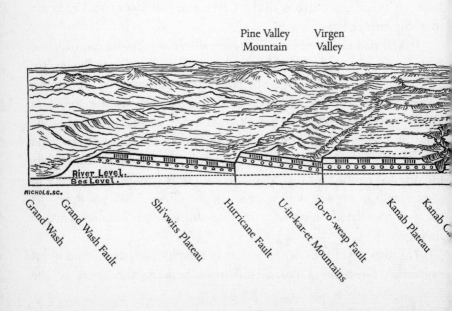

River Level.
Sea Level.

NICHOLS.SC.

Grand Wash

Grand Wash Fault

Shi'wits Plateau

Hurricane Fault

U-in-kar-et Mountains

To-ro'-weap Fault

Kanab Plateau

Kanab Ca

their mutually exclusive positions. But in the canyons through which the Powell expedition traveled, the rivers run so fast that they quickly sweep most of the potential sediment downstream. Bedrock is everywhere: it makes up the surrounding walls often half a mile to a mile high. Now and again, though more rarely than one would expect, it protrudes upward through the bed of the river. Alluvium is almost absent; the rafter sees only raw rock.

If one knows no better, the disparity between the width of rivers in a humid climate and the width of their valleys makes it easy to think of them as two independent entities. Take the lower Mississippi: the river itself may be a thousand feet or more wide, but, standing on one bluff, a person may not even be able to see the opposing bluff, tens of miles away. In such a setting, it may be natural to imagine that the wide valley came first and the much narrower river second. The Canyon of Lodore, where solid rock walls descend right down to the water, put the quietus

Paria Plateau

Kaibab Plateau · ...ab Fold · East Kaibab Fold · Marble Cañon · Echo Cliffs · Paria Fold · N. 80° E.

Figure 4. *Bird's-eye view of the Grand Cañon looking east from the Grand Wash. T artist whimsically used birds to indicate location: one bird, Echo Cliffs; two birds, Kaib Plateau; three birds, To-ro'-weap Cliffs; four birds, Hurricane Lodge; five birds, Shi-vw Plateau.*

on this simplistic notion. In a place like Lodore, one can no longer view the canyon and the river as separate entities. Here cause and effect inexorably marry river and valley. Powell's crew, short on education but long on common sense, likely would have scoffed at the notion that the canyons were older than the river. Only an eastern dude who had never seen a real western river, as they were beginning to, could entertain that idea. They could practically see the muddy river cutting the canyon. The self-taught Powell had managed to lodge himself in a river runner's hell but a geologist's heaven, a country where cause and effect were nakedly exposed.

In 1875, in response to a request from Congressman James Garfield of Ohio, his protégé John Wesley Powell published his geological conclusions as the three-part *Exploration of the Colorado River of the West and its Tributaries. Explored in 1869, 1870, 1871, and 1872*. In all his professional writings, Powell never showed, in Wallace Stegner's apt phrase, that "squidlike tendency to retreat, squirting ink," that typically marks the cautious scholar. To Powell, the lessons of the plateau country were so obvious that he only needed to write them down. They required little or no justification. He used scientific terminology sparingly and his language was eloquently simple. The periodic reports that he sent out to the *Chicago Tribune*, at the few places where that was possible, and his subsequent account in *Scribner's Magazine*, made exciting reading. Most of the scientific writing of Powell's day seems to us overly formal—ornate, long-winded, and pompous. Scholars writing in the century after him would fill their accounts with so many excrescent terms as to make their ideas incomprehensible to anyone but a specialist. Perhaps it was in part the era in which he wrote, but Powell was able to steer a course midway between pomposity and jargon.

The first section of his 1875 report was a popular account of the adventure. Together with his lectures and interviews, it launched not just his reputation but his celebrity. Though as its title indicates, the report covers more than the pioneering 1869 voyage, only from the title, not from the text, would a reader have any reason to suspect that beginning in 1871 and ending in 1872, Powell had made a second expedition from Green River, Wyoming, down through the canyons. The 1869 voyage had of necessity wound up focusing more on survival than science. A better-equipped second trip through now-known territory could carry out the scientific measurements that had eluded the first. But Powell's written account amalgamated the 1869 and 1871–72 trips into one, as though only the first occurred, an oddity that has vexed his biographers ever since. In a popular account, the blending of the two voyages might be barely excusable as poetic license, but in a scientific report it amounted to tampering with vital facts. Powell's failure in the 1875 report even to mention the names of the crew of the second voyage, brave men who risked their lives to accompany him, is inexcusable.

One of the members of the second crew was Frederick S. Dellenbaugh (1853–1935), an Ohio boy of seventeen who acted as artist and mapmaker. As the nineteenth century was drawing to a close, Dellenbaugh received Powell's blessing to write the missing account of the second trip. In the introduction to his *Romance of the Colorado River*, Dellenbaugh wrote that the men of the first voyage "had, under the Major's clear-sighted guidance and cool judgment, performed one of the distinguished feats of history." But he believed that "the men of the second party, who made the same journey, who mapped and explored the river and much of the country round-about, should have been accorded some recognition. The absence of this has sometimes been embarrassing for the reason that when statements of members of the second party were referred to the official report, their names were found to missing from the list. This inclined to produce an unfavorable impression concerning these individuals." In other words, Powell's omission had made the men of the second party appear no better than the great prevaricator, Risdon.

In 1895, Powell re-told the story in *Cañons of the Colorado*. His preface ended with these poignant words:

Many years have passed since the exploration, and those who were boys with me in the enterprise are—ah, most of them are dead, and the living are gray with age. Their bronzed, hardy, brave faces come before me as they appeared in the vigor of life; their lithe but powerful forms seem to move around me; and the memory of the men and their heroic deeds, the men and their generous acts, overwhelms me with a joy that seems almost a grief, for it starts a fountain of tears. I was a maimed man; my right arm was gone; and these brave men, these good men, never forgot it. In every danger my safety was their first care, and in every waking hour some kind service was rendered me, and they transfigured my misfortune into a boon.

To you—J. C. Sumner, William H. Dunn, W. H. Powell, G. Y. Bradley, O. G. Howland, Seneca Howland, Frank Goodman, W. R. Hawkins, and Andrew Hall—my noble and generous companions, dead and alive, I dedicate this book.

These were the men of the first voyage. Powell had expunged from the record the crew of the second, who performed the same services for him as the crew of the first, as though they never existed. To do so a second time, with some twenty years to have thought it over, adds insult to injury.

Powell's oversight constitutes one of the few, and strangely inexplicable, black marks on his career. Perhaps the second trip simply mattered far less to him than the first. Stegner thought so: "before he even started down it the second time, the river had lost much of its grip on Powell's restless imagination." Powell did leave the second voyage twice for extended periods, once to visit his pregnant wife in Salt Lake City, requiring his crew to complete the voyage themselves. Indeed, they ended it not at the Virgin, but more than one hundred river miles upstream at the mouth of Kanab Creek. Perhaps Powell had begun to lose interest in geology; perhaps he caught an early case of Potomac Fever. Or, as Worster points out, perhaps in Powell's mind there was no "second expedition," for it had started and stopped and proceeded with and without him, and ended short. At least he did acknowledge his later companions in his annual reports and by giving their names to landscape features as, for example, in Mount Dellen-baugh on the North Rim.

In the second part of Powell's report, entitled "On the Physical Features of the Valley of the Colorado," the self-taught backwoods scholar began to challenge the most eminent European geologists. But he did not enter into a figurative debate with them. Rather, as though the case presented by the Colorado Plateau brooked no debate, he simply stated his geological conclusions with little or no explanation. He began by noting that the topographic features of the valley of the Colorado are "in many respects unique, not reproduced, except to a very limited extent, on any other portion of the surface of the globe." He asked:

> To a person studying the physical geography of this country, without a knowledge of its geology, it would seem very strange that the river should cut through the mountains, when, apparently, it might have passed around them to the east, through valleys, for there are such along the north side of the Uintas, extending to the east, where mountains are degraded to hills, and, passing around these, there are other valleys, extending to the Green, on the south side of the range. Then, why did the river run through the mountain?

Buckland would have answered that the river found its way into a gorge cut by the great Flood of Noah. In the 1872 edition of his famous textbook, written three years after Powell's first voyage, Lyell waffled over whether rivers carve their own valleys. But neither of them, and indeed almost no one, had seen how the Colorado River had incised its way deep into the Colorado Plateau.

Two factors do briefly complicate the argument over the origin of river valleys. One is that during the recent Ice Ages, glaciers flowing downhill naturally found their way into pre-existing river valleys, goug-ing and scraping them out, leaving some modern valleys that are indeed too large for their streams. But after glaciers melt, they leave behind evidence of their passage in U-shaped valleys, in the moraines that drape the sides and mouths of the valleys, and in the sometimes-beautiful hanging-valley waterfalls, like Bridalveil Falls in Yosemite, the type example in the United States of a glaciated valley. The Colorado Plateau contains none of these glacial indicators. The other factor is that rivers are most able to erode where folding and faulting have prepared the way. Thus, where great faults break the surface, as in California and Scotland, and, it turns out, in parts of the Grand Canyon, valleys sometimes follow the trace of the fault plane on the surface. But that trace is almost always a straight line. A gorge eroded into rock along the trace of a fault does not meander back and forth and loop back on itself in horseshoe-shaped curlicues.

On the relationship between valleys and geologic structure, Powell quoted the views of A. R. Marvine, a geologist with the Geological and Geographic Survey of the Territories, who had published the year before Powell:

Meeting a softer bed a cañon will often have its course directed by it . . . but it soon breaks away and runs independently of the bedding. Many of the smaller ravines have had their positions determined by the structure; but in a broad sense the drainage is . . . independent of structure. Thus, while in places geological features may find expression in surface form, yet, as often, there may be no conceivable relation between topography and geology. The sub-aqueous [caused by running water] erosion, in smoothing all to a common level, destroys all former expression of geological character. . . .

For his part, Powell could see that the canyons changed direction constantly, usually bearing no relation to the rock structure and presenting no sign of following fault lines. Indeed, sometimes the drainage was in exactly the wrong place structurally:

From the foot of Red Canyon to the Gate of Lodore, a distance of thirty miles, the river runs through a valley known as Brown's Park, five or six miles wide, and enclosed by mountains. It is a curious fact that the central line of this valley corresponds to the axis of the fold; that is, had the fold been made, and left without erosion, the very summit would have been directly above the deepest part of the Park.

What Marvine and Powell were saying is this: If the earth were only a few thousand years old, the surface would be only slightly eroded and would retain its original shape. Anticlines, which are upfolds in the shape of an arch (∩), would show up as hills; their opposite, synclines (∪), would show up as valleys. Rivers would flow down the axes of synclines, between anticlinal ridges. But, as often as not, we find just the opposite: topographic valleys rest on structural anticlines and hills on structural synclines. Evidently, regardless of the underlying structure, a soft, easily eroded rock will almost always occupy a lower elevation than an adjacent resistant one. This alone demonstrates that erosion has been going on for a long time.

Powell was now ready to answer his obvious question: "Why did the river run through the mountain?":

The first explanation suggested is that it followed a previously formed fissure through the range; but very little examination will show that this explanation is unsatisfactory. The proof is abundant that the river cut its own channel. Again, the question returns to us, why did not the stream turn around this great obstruction, rather than pass through it? The answer is that the river had the right of way; in other words, it was running ere the mountains were formed; not before the rocks of which the mountains were composed, were deposited, but before the formations were folded, so as to make a mountain range.

Powell called these valleys, carved by streams older the landscape over which they now flow, *antecedent*. Exactly how does such a river come about?:

The contracting or shriveling of the earth causes the rocks near the surface to wrinkle or fold, and such a fold was started athwart the course of the river. Had it been suddenly formed, it would have been an obstruction sufficient to turn the water in a new course to the east [referring to the Uinta Range], beyond the extension of the wrinkle; but the emergence of the fold above the general surface of the country was little or no faster than the progress of the corrasion [abrasion] of the channel. We may say, then, that the river did not cut its way *down* through the mountains, from a height of many thousands of feet above its present site, but, having an elevation differing but little, perhaps, from what it now has, as the fold was lifted, it cleared away the obstruction by cutting a cañon, and the walls were thus elevated on either side. The river preserved its level, but mountains were lifted up; as the saw revolves on a fixed pivot, while the log through which it cuts is moved along. The river was the saw which cut the mountains in two.

But Powell could not entirely suppress his romantic side:

The upheaval was not marked by a great convulsion, for the lifting of the rocks was so slow that the rains removed the sandstones almost as fast as they came up. The mountains were not thrust up as peaks, but a great block was slowly lifted, and from this the mountains were carved by clouds—patient artists, who take what time may be necessary for their work. We speak of mountains forming clouds at their tops; the clouds have formed the mountains.

He was not the first to hit upon the concept of antecedence. Indeed, the history of science shows that many, if not most, discoveries happen when one of several people hot on a trail manages to get there first. Often the new idea has been in the scientific air and more than one person breathes it. By 1857, the year Powell entered Oberlin College, alumnus Ferdinand Hayden had already made a substantial name for himself as a fossil and mineral collector. Indeed, he was the first geologist to study large regions of the West, including what would become Yellowstone National Park. In describing the Rocky Mountains, Hayden showed that he clearly grasped the idea of antecedence:

Another illustration of the gradual and long continued rise of the country may be found in the immense chasms or cañons which have been formed by the streams along the mountain sides. We can only account for them on the supposition that as the anticlinal crest was slowly emerging from the sea, the myriad sources of our great rivers were seeking their natural channels, and that these branches or tributaries began their erosive action long before the great thoroughfares, the valleys of the Mississippi and Missouri, were marked out. The erosion would go on as the mountains continued slowly rising at an almost imperceptible rate, and in the process of time the stupendous channels which everywhere meet us along the immediate sides of the mountains would be formed.

Throughout his career, Powell was prone to invent names for whatever he studied, whether the field was geology, philosophy, or psychology. Most of his names disappeared at once, but a few caught on. Perhaps if Hayden had given the concept that he described above a name, he and not Powell would have received credit for antecedence.

Powell defined two other kinds of rivers. An original stream that follows the slope of the topography or rock structure, he called *consequent*. Any stream must start out as consequent, but, if it survives, must become either antecedent or *superposed*, the third possibility. In an antecedent valley, the land rises beneath a pre-existing river, as he believed the Uinta Range had done. If the river is to maintain its course, it must cut down through the incipient obstruction. More of its energy then goes toward cutting vertically than laterally, causing the river to slice a canyon right into the rising obstacle. In a superposed (sometimes called superimposed) valley, almost the opposite happens. Layers of younger sediment bury a geologic structure, perhaps an old mountain range. Later, a consequent river begins to erode these sediments. Gradually, erosion lowers the elevation of the surface, perhaps by many thousands of feet, until the river impinges upon and strikes the buried older structure. If that structure is made of harder rock and yet the river is to maintain its course, it must carve a canyon through the structure. Although Powell briefly considered and rejected the idea, the Green River, instead of being antecedent, could have lowered itself gradually through the overlying sediments until it struck the buried Uinta structure, into which it cut the Canyon of Lodore. He did little more than assert that the Green is antecedent; as subsequent work would show, that was not enough.

Whether a valley is antecedent or superposed is a distinction with a difference. If the Green is antecedent, it is older than the Uinta Mountains. If the entire Colorado River is antecedent, it is older than the Colorado Plateau. Both rivers then could be far older than the structures they cross: they could be truly ancient. But if the rivers are superposed—let down onto the structures they now cut—they are younger than those structures, and could be a great deal younger. If rivers can be around for scores of millions of years, they have the deep time of geology to work their will on landscapes. Later we will see that geologist William Morris Davis proposed a model of landscape evolution in which a land surface remains in one place for a long time, more or less passively submitting to the streams that gradually dissect it. But if rivers are relatively young, yet still they have accomplished vast erosion, as in the Uintas and the Colorado Plateau, some other factor must be at work. Something else must enhance a river's erosional power.

Powell believed that the Colorado River is antecedent all the way from the Gate of Lodore to the Virgin River. Like every geologist who has studied the Grand Canyon, Powell noted near the Toroweap section one of its most visible and important features: two giant, near-vertical faults that have broken and offset the rocks. Upriver, rock layers have sometimes been bent into equally large step-folds, which Powell named monoclines. These displacements extend for scores of miles north and south of the canyon. A crucial kind of evidence that geologists were to spend decades addressing was whether the Colorado River cut across these structures, or at least in part bent around them. A truly antecedent river would incise its way downward, oblivious to the underlying geologic structure. But a superposed one, descending and caught unawares, might be more apt to defer to structure. Figure 3, page 48, taken from Powell's 1875 report, illustrates the major faults and folds of the Grand Canyon.

Powell left no doubt where he stood on whether the river or the structures were older:

> Though the entire region has been folded and faulted on a grand scale, these displacements have never determined the course of the streams. All the facts concerning the relation of the water-ways of this region to the mountains, hills, canyons, and cliffs, lead to the inevitable conclusion that the system of drainage was determined antecedent to the faulting, and folding, and erosion, which are observed, and antecedent, also, to the formation of the eruptive beds and cones.

IN THE MANY HUNDREDS OF MILES Powell had logged canoeing the rivers of the Midwest, he had countless opportunities to confirm Playfair's observation that tributary streams have "such a nice adjustment of their declivities, that none of them join the principal valley, either on too high or too low a level." Surely he had come to agree with Playfair that this "circumstance would be infinitely improbable if each of these valleys were not the work of the stream that flows in it."

Let us remember that during the 1860s and 1870s, and even later, leading geologists on both sides of the Atlantic still disagreed with Playfair. Both Charles Lyell and Clarence King were unwilling to attribute valleys entirely to the work of the rivers that flow in them, even though King had seen the American West. But if rivers did not carve their own valleys, what did? The geologists of the second half of the nineteenth century did not know and had to appeal to erosion by ocean waves and currents, after which the land rose above sea level. This unsatisfactory theory left geologists unable to explain the origin of river valleys, one of the most fundamental and ubiquitous features of the earth's surface.

That failure in turn gave the floor to religious fundamentalists, who could claim that the waters of the Flood carved the valleys.

If rivers do not carve their valleys, many nearly universal observations remain inexplicable. Why then do rivers and their valleys both grow larger downstream? Why do tributaries enter at the level of the main stem, as Playfair observed? Why do tributaries usually enter in such a way as to make an acute angle pointing downstream? Why do small streams and small valleys enter large streams and large valleys, rather than the other way round? Why do streams and tributaries often form complex, branching, "dendritic" patterns? And so forth. But if rivers carve their own valleys, all these observations make sense.

Powell did more than corroborate Playfair. He used the ability of streams to carve their valleys to work back to the history of rivers, valleys, and landscapes. Without this essential breakthrough, scientists could never have conceived whole branches of earth science, such as physical geology and geomorphology.

Again, we have to ponder how much Powell's insights owed to his innate perspicacity and how much to the country in which he traveled. His ideas may have begun to take shape on the Ohio or the Mississippi, well before he got to the white water of the Colorado, but he never said so. Had he remained teaching geology at a small college on the glaciated plains of the Midwest, it might have taken a stroke of genius for him to understand how rivers and valleys relate. But floating down the Green and Colorado and seeing the many narrow slot tributary canyons, each fed by a much smaller, intermittent stream, yet having carved its valley as deeply as the main stem, he may have been no more able to escape his conclusions than his "granite prison."

One of Powell's first opportunities to observe a large tributary entering the Colorado river system was far above the Grand Canyon, on the Green just below the Canyon of Lodore, where the Yampa, flowing in from the east, joins at Echo Park in present-day Dinosaur National Monument. At this beautiful spot, some of his men claimed they could hear as many as ten or twelve successive echoes bouncing back and forth from the high, facing canyon walls. And well they may have, for a modern traveler without the most acute hearing can detect five or six.

According to the account of one of his crew, Echo Park almost ended Powell's voyage—indeed, his life—for during one of his almost daily climbs, he found himself stranded on a rock precipice sixty to eighty feet high, unable to go up or down. In what became perhaps his most famous escapade, his climbing companion had to rescue Powell by converting his own long johns, already surely in a disreputable state, into a rope and drawing the one-armed alpinist up to safety. A more authoritative account says this episode took place three weeks later and well downstream, but wherever the underwear saved the day, the story remains a great one. From Mount Dawes, high above the Yampa, through the crystal air of a time before the haze from giant, coal-fired power plants polluted the west, Powell could see north more than 100 miles to the Wind River range and east to the western slopes of the Rockies, more than 150 miles away, where lay the headwaters of the Yampa.

The farther up a valley like the Yampa one travels, the steeper the terrain and the faster the river flows. The faster a river flows, the more it erodes. Conversely, as the gradient of the Yampa lessens downstream, so does its ability to erode. Where it enters the Green, the two flow over the same slope and therefore have the same velocity. At the junction, they simply merge together, effortlessly becoming one, with no sense that one stream has joined the other. One of Powell's men, Sergeant George Bradley, expressed surprise that even downstream from Echo Park, where the Green and former Grand join to make the mighty Colorado, the merger is "so calm, wide, and quiet."

It must have been in observing the seamless merger of so many tributaries and main stems that Powell got the spark for the second of his two great contributions to geology. He understood that the Yampa continuously erodes its valley down toward the level of the Green. Deeper than that the Yampa cannot cut, for then the waters of the Green would flow back up the Yampa, uphill. Conversely, were the Green to encounter softer rock and lower its channel faster, near the junction of the two the Yampa would also acquire a steeper gradient and therefore would begin to erode faster, eventually erasing the steeper gradient and regaining its level entry. Thus, erosion has forever locked together the Yampa and the Green—and any tributary and main stem. For the Yampa, the level of the Green represents an inescapable but always receding ultimate destination, one that Powell called "base level."

Echo Park is one of the most monumental and beautiful scenes in the West. With the six-shooters of Butch, Sundance, and their compadres long since holstered, danger seems far away. Not so the Grand Canyon—there the many narrow slot canyons threaten danger almost literally around every corner. And there the power of the concept of base level becomes evident. Consider Kanab Canyon, one of the largest valleys to enter the Grand Canyon, and far too large to be called a "slot" canyon. It comes down from the north to meet the Colorado at river mile 144 below Lee's Ferry, the way geologists and river runners measure distance through the Grand Canyon (see Figure 8, page 188). Were it not for the contrast with the Grand Canyon, we would consider Kanab Canyon one of the major features of the Colorado Plateau. Powell knew it well, for he visited it in 1870 and used it as his exit route when he ended the second voyage in 1872.

Rafting down the Colorado, Kanab Canyon remains hidden until one is almost upon it. Then what appeared to be merely the continuation of the wall of the Grand Canyon reveals itself instead as a nearly vertical, mile-high, but relatively thin, canyon in its own right. Seeing Kanab Canyon and the many other even thinner, true slot canyons that enter the Grand Canyon, Powell must have asked himself why most if not all of the streams that eroded them did not simply wander across the plateau and then plunge down from the rim in waterfalls thousands of feet high. How is it that the narrow, slot canyons with their much smaller streams, manage to so nicely adjust their declivities that they enter at the level of the main stem Colorado?

To make the question even more difficult, Kanab Creek, like most of the side canyons, and in contrast to the perennial Colorado, is usually dry. Flash floods account for the removal of all the solid rock that once filled Kanab Canyon. Like much in geology, this proposition at first seems counterintuitive. How could a stream that contains water only a few times a year, and then for only hours or days, erode its valley to the same depth as the Grand Canyon, carved by one of the most persistent rivers on the continent?

The answer derives from applying the basic laws of physics *over geologic time*. As the Colorado in the Grand Canyon lowers its channel inexorably downward, it continually lowers the base level and steepens the gradients of all the streams that enter it. These tributaries gain more erosive power and in turn cut their channels downward more rapidly, approaching but never quite reaching their base level: the continually falling Colorado River. The more rapidly the Colorado drops, the faster the tributaries must erode in their inevitable quest to keep pace. Where gradients are steep and little soil and vegetation holds back the rainwater, streams that flow only now and then evidently can cut down effectively enough to keep up with a vastly larger perennial master stream. They have done so.

That is the technical answer. But looking from the Colorado River up Kanab Creek or, even more so, up one of the narrow slot canyons, even the hard-headed geologist may sense some powerful and mysterious force, some manifest destiny, that has sent the intermittent streams that carved the side canyons questing ever downward, probing for but never quite reaching the level of the much more voluminous, persistent, formerly silt-laden, and rapidly dropping Colorado, a process that allows, or rather requires, tributaries fed by sporadic local thunderstorms to keep up with a giant main stem fed by snow melting a thousand miles away. The power of the concept of base level now becomes evident. Even an intermittent stream inherits the strength of ten, enough to slice through a mile of solid rock, like a David fated for eternity to strive to match Goliath and, miraculously, doing so.

The true slot canyons are much smaller than Kanab Canyon: narrow, twisting, and sometimes delicate niches. Sunlight suffusing through them can illuminate their colorful rocks and render them as beautiful as any cathedral. But a cathedral is a sanctuary, a refuge from the world's trials. The slot canyons are among the most treacherous natural phenomena, where heaven can turn into hell in an instant.

One of the most accessible and beautiful is Antelope Canyon, which once entered the Colorado at Glen Canyon, but now flows into Lake Powell a few miles above Page, Arizona. In some spots, the canyon is so narrow that one's extended arms can touch each side at once. Understandably, Antelope Canyon is a favorite destination for tour groups and photographers. On August 12, 1997, during an El Niño western summer, an outdoor travel organization called Trek America led a group of international tourists to Antelope Canyon. Authorities had warned of the possibility of a severe thunderstorm, a frequent occurrence during the northern Arizona summer, as Powell and his men had learned the hard way. According to the Chief of Police of the Navajo Nation, authorities had warned the tour group that a flash flood could occur. But at Antelope Canyon only a trace of rain fell and all appeared to be well. The Trek America group had already toured the canyon, but several wanted to return to use the rest of their film. What they did not know was that forty-five minutes earlier, fifteen miles away and 2,000 feet higher, out of sight and hearing of anyone in the narrow canyon, the severe thunderstorm warning had come true. An inch-and-a-half of rain fell, half of it in fifteen minutes. With no soil or vegetation to soak it up, in the "slickrock" country of the Colorado Plateau rainwater runs nearly unchecked across the surface and down into drainage channels, where it quickly coalesces and magnifies. Within minutes, a dry arroyo can fill with a wall of water higher than a person's head and traveling much faster than a person can run.

Without warning, an eleven-foot wall of muddy, debris-choked water rushed down Antelope Canyon and engulfed the tourists at the bottom, leaving the others to look on horrified from higher ground. The only survivor was the Trek America guide, who told the investigator how he had tried to anchor his feet and hold two others fast, but lost his own grip and washed downstream, winding up dazed, bruised from head to toe, blinded by silt, and stripped naked. Eleven tourists, including seven from the Trek America party, died. Two of the bodies were never recovered. To add to the tragedy, among the missing were the parents of two children who had remained behind.

Flash floods are inevitable in canyon country—they created most of the canyons. In the same week as the Antelope Canyon flood, 100 miles downstream a wall of water rushed through the Havasupai Indian Reservation, requiring a helicopter to airlift residents and tourists. Another 100 miles southwest of there, also in the same week, an Amtrak train derailed after a flash flood and damaged a bridge, injuring more than 150 people. One who has witnessed a flash flood on the Colorado Plateau has no trouble believing that such an event, repeated through the ages, can carve a deep canyon.

THE CONCEPT OF BASE LEVEL APPLIES not only to the tributary streams that directly enter their main stem, but to a river that enters the ocean. Powell concluded that sea level was the "grand base level, below which the dry lands cannot be eroded." He understood that base level was a theoretical plane that no river could ever attain, for "the action of a stream in wearing its channel ceases . . . before its bed has quite reached the level of the lower end of the stream."

But following the implications of base level as far as we can leads to other insights. As the Colorado erodes and deepens its channel, heading for its destination—sea level at the Gulf of California—it lowers the base level of its tributaries concomitantly. As they descend in lockstep with the Colorado, the base level of all the streams that enter them—the second-order set of tributaries—lowers in turn. Then the third-order set lowers, and so on. Extend the idea to all the drainages of the Colorado Plateau, and indeed to all streams of whatever size everywhere, and the concept emerges of a grand, universal reduction of all landscapes toward a common, low surface, approaching sea level asymptotically but never reaching it.

Fair enough, but then the fundamental question raised earlier reappears: because streams could have reduced the continents to sea level in only about twenty-five million years at their average rate of erosion, why have they not done so? Powell and the other geologists of his day recognized that because the earth's surface has not been reduced to a common, low elevation, some unknown force, some powerful agent, must repeatedly uplift regions of continents, bringing them higher above their base levels, rejuvenating their streams, and renewing the cycle of erosion. In Powell's day the ultimate cause of these "upheavals" was one of the great mysteries of geology. To solve it would require a century.

The gods condemned Sisyphus to roll a stone to the top of a mountain, only to have it roll down, requiring him to start over, for Eternity. Erosion does the opposite: it lowers mountains, only to have plate tectonics heave them up again, starting the process anew. And a good thing it is, too, for if all the surface areas of the earth had been brought to a common level, our planet would be covered with half a mile of sea water and we would not exist. Plate tectonics is not only the ultimate cause of even the majestic Grand Canyon—it is the root cause of geological activity and therefore, in a sense, of life itself.

But such notions lay far in the future, well beyond the career of the young, self-taught Midwestern professor, brave adventurer, and one-armed mountain climber. Nevertheless, the ending of Powell's 1875 report hints that he just might have had an intuition of how far the concept of base level might extend (and shows how quick he was to accept Darwin's evolution): "Thus ever the land and sea are changing; old lands are buried, and new lands are born, and with advancing periods new complexities of rock are found; new complexities of life evolved."

In John Wesley Powell, the man and the country had come together, as poet Sam Walter Foss had hoped:

> Bring me men to match my mountains,
> Bring me men to match my plains,
> Men with empires in their purpose,
> And new eras in their brains.

But Powell was not the first geologist to see the canyon country. To put him in context, we need to understand the antecedents of his explorations, which began more than three centuries before him. Then we can return to see how the Major's tersely expressed and barely evidenced theory of antecedence fared with his fellow geologists.

Chapter 5

Seven Cities of Gold

~~~~~~

Although John Wesley Powell and his crew were the first to make it all the way through the Green and Colorado canyons, white men had reached the rim of the Grand Canyon more than three centuries before. By the 1520s, with the Pilgrims' landing at Plymouth Rock still a hundred years in the future, the Spanish conquest of the Western Hemisphere was well underway. The conquistadors who had stepped with Pizarro across his line in the sand had vanquished the Incas of Atahuallpa and claimed for Spain vast riches of gold and silver. To the north, Cortés had outmaneuvered Montezuma, who mistook him for the god Quetzalcoatl, and won control of Mexico and Aztec treasure. But the Royal appetite for precious metal was not to be slaked—on gold and silver rode the fortunes of men and nations.

In 1530, the Viceroy of New Spain, Nuño de Guzman, heard an intriguing tale from an Indian the Spaniards called Tejo. He said he was the son of a trader who, when Tejo was a boy, had traveled into the backcountry to trade feathers but returned with a trove of gold and silver. On subsequent trips, Tejo had journeyed with his father and seen seven large towns with streets lined with metal workers. The story was not hard for a Spaniard to believe. The Incas had shown that native peoples could extract precious metals and work them skillfully; after seeing the great Aztec city of Tenochtitlan, what wise administrator would say that others like it might not exist? Guzman assembled an army of several hundred Spaniards and thousands of Indians and set off in search of the Seven Cities. They reached the Mexican outpost of Culiacán in present-day Sinaloa but failed to find the fabulous cities. Perhaps they did not exist, or, the searchers must have fervently hoped, more and better information would serve to locate them.

In 1536, near Culiacán, a group of Spaniards on a slaving expedition were dumbstruck to see, approaching them out of the wilderness, three rag-tag Spaniards, an African, and a band of Indians. The leader told a tale that was as odd as his name: Álvar Núñez Cabeza de Vaca ("Cow's Head"). In the eleventh century, to honor an ancestor who had marked an unguarded pass with the skull of a cow and saved a battle for the Christians, the King of Navarra had given the surname to the man's matrilineal line. The Cabeza de Vaca (c.1490–c.1557) who emerged from the jungle had been treasurer of the expedition of Panfilo de Narvaez, who had won a royal patent to found a colony in Florida. In April 1528, Narvaez landed on the west coast of Florida and elected to send his ships ahead while he and 300 men reconnoitered, planning to have the two parties meet up farther along the coast. But once ashore, the men never again saw the ships. The Apalachee Indians of North Florida gave the Spaniards food and shelter, but expelled them after they took the chief hostage, reducing the Conquistadors to hiding in a swamp and eating their horses. To escape this purgatory, they constructed rafts of timber and horsehide and set sail, bound back for Cuba.

Instead of reaching that island, a storm swept the eighty remaining men ashore near Galveston, Texas in November 1528. Again, the Spaniards soon squandered the friendship of the Indians, for according to Cabeza de Vaca, "half the natives died from a disease of the bowels and blamed us." Nevertheless, the dwindling number of Spaniards managed to live for four years among the Indians, Cabeza de Vaca becoming a trader and medicine man. By 1532, only he, the African Estevanico, and two other Spaniards remained alive. Desperate, they decided to head west to try to reach the Spanish settlements in Mexico. In one of the epic journeys of history, the four traveled across the southwest and well down into Mexico, passing from village to village, often trailed by bands of Indian well-wishers. In July 1536, eight years and several thousand miles after they landed in Florida, Cabeza de Vaca, Estevanico, and the other two reached Culiacán. As Cabeza de Vaca remembered, his countrymen were "dumbfounded at the sight of me, strangely dressed and in company with Indians. They just stood staring for a long time."

Cabeza de Vaca told of the strange and wonderful tribes he had met, some of whom lived in "powerful villages, four and five stories high, of which they had heard a great deal in the countries they had crossed, and other things very different from what turned out to be the truth." He had seen "streets lined with goldsmith shops, houses of many stories, and doorways studded with emeralds and turquoise." By this time, Antonio de Mendoza had succeeded Guzman. Cabeza's affirmation of what Mendoza wanted to believe was more than enough to persuade him to send out a trusted man, Friar Marcos de Niza, to locate the Seven Cities, for surely it was they that Cabeza de Vaca described. Accompanied by Estevanico, who soon perished at the hands of Indians, having shown too much interest in their turquoise and their women, Marcos traveled north for hundreds of miles. Finally, historians believe, Marcos reached a group of villages called Cibola—surely the fabled Seven Cities. Actually, they were the Zuni settlements.

The report that Marcos gave upon his return apparently inspired Mendoza, who must have been savoring another Inca or Aztec empire to conquer, to appoint new expeditions. One party, under Hernando de Alarcón, sailed north up the coast to the mouth of a large river, which turned out to be the Colorado. Another, led by Francicso Vazquez de Coronado (1510–1554), marched overland from Compestela in New Galicia in 1542 with a large force of soldiers, Indians, stock animals, and priests, including Friar Marcos. The expedition reached Cibola but instead of treasure found only resistant Indians, who came close to killing Coronado. Pedro de Castenada, the chronicler of the Coronado expedition, wrote that the Spaniards were more than disappointed: "When they saw the first village, which was Cibola, such were the curses that some hurled at Friar Marcos that I pray to God He may protect him from them."

But in the search for treasure, great riches always seem to lie just over the horizon, at the end of the rainbow. Coronado sent twenty men under the command of Pedro de Tovar, still farther north. Tovar reached a place called Tusayan in July, but again found no gold. He did return with a bit of turquoise and a story from the Tusayan elders, who told of a great river to the west where lived "some people with very large bodies." Coronado dispatched Don Garcia Lopez de Cardenas and a dozen men to find that river. A few day's march should have taken them to the edge of Marble Canyon, near present day Lee's Ferry, yet the trip took twenty days, leading historians to suspect that the Indian guides, having had their fill of Conquistadors, had sent them on a wild-goose chase. In any event, according to Castenada, Cardenas and his men eventually reached the great river:

He [Cardenas] was well received when he reached Tusayan and was entertained by the natives, who gave him guides for his journey. They started from there loaded with provisions, for they had to go through a desert country before reaching the inhabited region, which the Indians said was more than twenty days journey. After they had gone twenty days, they came to the banks of the river. It seemed to be more than 3 or 4 leagues in an air line across to the other bank of the stream which flowed between them. They spent three days on this bank, looking for a passage down to the river, which looked from above as if the water was 6 feet across, although the Indians said it was half a league wide. It was impossible to descend, for after these three days, Captain Melgosa and one Juan Galeras and another companion, who were the three lightest and most agile men, made an attempt to go down at the least difficult place and went down until those who were above them were unable to keep sight of them. They returned about 4 o'clock in the afternoon, not having succeeded in reaching the bottom on account of the great difficulties which they found, because what seemed to be easy from above was not so, but instead was very hard and difficult. They said that they had been down about a third of the way and that the river seemed very large from the place which they reached, and that from what they saw, they thought the Indians had given the width correctly. Those who stayed above had estimated that some huge rocks on the sides of the cliffs seemed to be about as tall as a man, but those who went down swore that when they reached these rocks, they were bigger than the great tower of Seville. They did not go further up the river, because they could not get water.

Having found more empty space than gold, Cardenas returned to the Zuni Pueblos, only to find the Indians about to revolt over the Spaniards' duplicity. To demonstrate the fate of those who dared challenge him, Coronado ordered 200 Zunis captured and burned alive. The Indians tried to escape, but "the horsemen gave chase [and] as the country here is level, not a man of them remained alive."

Coronado, believing the tale of a swarthy Indian slave they had nicknamed "The Turk," next set off in search of yet another great city, this one said to be far to the northeast in a place called Quivira. After crossing New Mexico, northern Texas, Oklahoma, and much of Kansas, to reach a point about 2,500 miles from where his expedition had started, Coronado found only "a very brutish people, without any decency whatever. In their homes or in anything." The Spaniards asked The Turk why he had lied and had guided them so far out of their way. He said that "his country was in that direction and that, besides this, the people at Cicuye [where the Spaniards first met The Turk] had asked him to lead them off on to the plains and lose them, so that the horses would die when their provisions gave out, and they would be so weak if they ever returned that they would be killed without any trouble, and thus they could take revenge for what had been done to them."

Coronado, in his report to Mendoza, put it this way:

I remained twenty-five days in this province of Quivira, so as to see and explore the country and also to find out whether there was anything beyond which could be of service to Your Majesty, because the guides who had brought me had given me an account of other provinces beyond this. And what I am sure of is that there is not any gold nor any other metal in all that country, and the other things of which they had told me are nothing but little villages, and in many of these they do not plant anything and do not have any houses except of skins and sticks, and they wander around with the cows; so that the account they gave me was false, because they wanted to persuade me to go there with the whole force, believing that as the way was through such uninhabited deserts, and from the lack of water, they would get us where we and our horses would die of hunger. And the guides confessed this, and said they had done it by the advice and orders of the natives of these provinces.

To get ahead of our story, contrast this bleak but honest assessment with that of the boosters of western expansion three centuries later in a testimonial in *Western Trail*, the Rock Island Railroad's gazette: "Why immigrate to Kansas? Because it is the garden spot of the world. Because it will grow anything that any other country will grow, and with less work. Because it rains here more than in any other place, and at just the right time."

And the fate of the duplicitous Turk? "They garroted him." With nothing to show and short of supplies, Coronado returned to Zuni, where three of his priests asked to stay behind to found a mission. No sooner had Coronado departed than the Zunis killed the three Padres. Somewhere in this long traverse, Coronado's horse stepped on his head.

After the expedition, the authorities indicted Coronado for his failures, but the court acquitted him. Though a subsequent inquiry brought him a fine and the loss of the Indians on his estate, Coronado did retain for life his seat on the Council of Mexico City. The authorities also arrested Cardenas and sent him back to Spain. Yet in only two years, the two explorers claimed more territory for Spain than the Roman Empire acquired in five centuries. But in the eyes of their masters, scenery, and even land itself, was valueless in comparison to treasure, leading Spain to cede all the territory Coronado had won. His men and his superiors could no more estimate and appreciate the Grand Canyon than they could understand the Indians they slaughtered. The cultural gap, like the Canyon, was too wide. Ironically, Coronado must have passed close by two great gold and silver lodes that prospectors would later discover in

Arizona. Had he made a detour to them, rather than to the Grand Canyon, world history might have turned out differently. Even so, Coronado did leave a different kind of legacy: some of the horses he lost wound up in the hands of the Apache and the Comanche, who used them to become among the finest cavalry the world has known.

In 1604, an expedition led by Juan de Oñate, Governor of New Mexico, passed quickly by the Grand Canyon, pausing to name the river they found there el Rîo Colorado, "the Red River." But it turned out to be the Little Colorado, not the main stem. With no reason to return, more than two hundred and fifty years elapsed before another Spaniard probed the Grand Canyon and the Colorado River. In the mid-1700s, Friar Francisco Garcés, one of the most intrepid of the Spanish explorers, undertook five great treks from his mission at San Xavier de Bac, then in Mexico but now in Arizona. On one journey, he followed the Gila down to the Colorado and may have descended to its mouth. Garcés accompanied De Anza on his successful 1774 expedition to find a route connecting the missions in Mexico and those in California. In 1776, on his way to the Hopi Villages, Garcés visited Havasu Canyon, the northern extension of Cataract Creek (see Figure 8, page 188), and spent several pleasant days with the Havasupai Indians. He climbed back to the plateau and journeyed east, seeing the Grand Canyon, the Kaibab Plateau, and the junction of the Colorado and the Little Colorado at Marble Canyon. He reached the Hopi Village of Oraibi in July. But two days later, the fourth of July 1776, the Hopis expelled him. Three years after that, Yuma Indians killed Garcés.

In the same year in which Garcés saw the Grand Canyon, another party of Spaniards sought a new route from Santa Fe to the California missions. Fathers Escalante and Dominguez traveled well east of the Grand Canyon and the demonstrably unfriendly Indians, planning to go north to the latitude of Monterey, California, and from there to strike due west to the Pacific. On July 29, 1776, their party left Santa Fe; by mid-August they had reached the Dolores River in western Colorado. There they became lost, saved only by friendly Indians who set them on the right route. The Spaniards headed west as planned, crossed the Green River, and by October had reached the vicinity of present-day

Cedar City, Utah. With winter approaching and unknown mountains and deserts between them and the Pacific, they put their fate in the hands of God and cast lots to see whether to continue or to return to Santa Fe. Fortunately, returning won. But retracing their circuitous route would have taken far too long, had they been able to find it again. Having heard rumors of a place where they might ford the Colorado, they crossed the Virgin River near present-day Toquerville, Utah, and headed east, eventually traversing the Kaibab Plateau and striking the Paria River. An eleven day march down the Paria brought them to the Colorado at the site of Lee's Ferry, but they found they could not ford. They named the place San Benito—a monk's robe of penance—and added the sobriquet Salsipuedes, which translates to: "Get out if you can." Finally they found a ford at Wahweep Creek, thirty-five miles above the mouth of the Paria. There, at Ute Ford, as the Indians called it, or El Vado de los Padres, as it become known to the Spaniards, they cut steps in the sandstone cliffs, crossed the Colorado River, and went on to reach Santa Fe on January 2, 1777. For a century after Escalante and Dominguez, through a stretch of river several hundred miles long, El Vado provided the only way to ford the Colorado. Today, like beautiful Glen Canyon itself, the Crossing of the Fathers lies under hundreds of feet of lake water.

Garcés, the last of the Spanish explorers to visit the Grand Canyon, preceded the first American to do so by eighty years. Garcés wrote of having to reach the Havasupai Tribe, who lived below the canyon rim, by climbing down a ladder at the point where springs emerge to feed Havasu Creek. Four decades later, the same ladder, or one of its descendants, greeted members of the next expedition. An American Army Lieutenant, Joseph Christmas Ives (1829–1868), led the party, accompanied by a hefty German artist, Baron Frederick W. von Egloffstein:

> We were deeper in the bowels of the earth than we had ever been before, and surrounded by walls and towers of such imposing dimensions that it would be useless to attempt describing them; but the effects of magnitude had begun to pall, and the walk from the foot of the precipice was monotonously dull; no sign of life could be

discerned above or below. At the end of thirteen miles from the precipice an obstacle presented itself that there seemed to be no possibility of overcoming. A stone slab, reaching from one side of the cañon to the other, terminated the plane which we were descending. Looking over the edge it appeared that the next level was forty feet below. This time there was no trail along the side bluffs, for these were smooth and perpendicular. A spring of water rose from the bed of the cañon not far above, and trickled over the ledge, forming a pretty cascade. It was supposed that the Indians must have come to this point merely to procure water, but this theory was not altogether satisfactory, and we sat down upon the rocks to discuss the matter.

Mr. Egloffstein lay down by the side of the creek, and projecting his head over the ledge to watch the cascade, discovered a solution of the mystery. Below the shelving rock, and hidden by it and the fall, stood a crazy looking ladder, made of rough sticks bound together with thongs of bark. It was almost perpendicular, and rested upon a bed of angular stones. The rounds had become rotten from the incessant flow of water. Mr. Egloffstein, anxious to have the first view of what was below, scrambled over the ledge and got his feet upon the upper round. Being a solid weight, he was too much for the insecure fabric, which commenced giving way. One side fortunately stood firm, and holding on to this with a tight grip, he made a precipitate descent. The other side and all the rounds broke loose and accompanied him to the bottom in a general crash, effectively cutting off the communication.

After a look around, the party confronted the task of retrieving the portly painter. This they did by making a rope of the slings from the soldiers' muskets. "Whether it would support his weight was a matter of experiment. The general impression was that it would not, but of the two evils—breaking his neck or remaining among the [Indians]—he preferred the former."

IVES'S EXPEDITION WAS ONE OF several mounted by Congress in the mid-1800s to seek out the best route for the Transcontinental Railroad that the nation desperately desired. In 1853, four years before the Ives expedition, Army Lieutenant Amiel Weeks Whipple (1817–1863) had led his party west along the 35th parallel from Fort Smith, Arkansas. They reached the desert southwest, only to spend three futile months attempting to blaze a trail from the Zuni Pueblo to the Colorado River. They passed within a few miles of the South Rim of the Grand Canyon and, though the way was open, kept right on going, "even across terrain that offered not the slightest impediment to travel." But perhaps the Whipple party can be forgiven, for they were far from the only travelers to pass close by, unaware that the Grand Canyon lay only a short distance away. As any visitor knows, the canyon remains almost invisible until one is right on top of it.

Congress directed Ives, who had served with the Whipple party, in the winter of 1857–1858 to navigate as far as he could up from the mouth of the Colorado River. Over the years, steamboats had sailed up the Colorado as far as Fort Yuma, but the War Department wanted Ives to go farther upstream to see whether the river was navigable closer to the remote military bases in New Mexico and Utah that needed supplying. Ives might have chartered the Colorado River steamboat of one Captain Alonzo Johnson, who already plied up and down the river. Instead, Ives ordered his own steamer built in Philadelphia and disassembled, shipped in pieces through the Isthmus of Panama to San Francisco, and from there to the mouth of the Colorado River, where it was reassembled. His explanation for having gone around Robin Hood's Barn was that Johnson's company "was unable to spare a boat, except for a compensation beyond the limits of the appropriation."

Ives got his long-traveled vessel as far as the mouth of Black Canyon, below present-day Las Vegas, whereupon he hit a rock, ending the boat trip. In the next century, engineers would remove this same rock to make way for Hoover Dam and Lake Mead. Ives then divided his party and marched with one group upriver, along the route of Garcés. They ascended the Grand Wash Cliffs that mark the western edge of the Colorado Plateau and the exit of the Grand Canyon (see Figure 8), to the plateaus above the South Rim. From there they descended via dry Peach Springs Canyon (also shown in Figure 8) to the junction of Diamond Creek and the Colorado, along the way drawing up the stranded von Egloffstein, just as the Major would later be drawn up by a pair of long johns. From the river's edge, they ascended and crossed the plateau eastward past the San Francisco Peaks, north of present-day Flagstaff, detoured to the Hopi Villages, and went on to muster out at Fort Defiance, Arizona. In his report, Ives made one of the least accurate predictions in history:

> It [the Grand Canyon] looks like the Gates of Hell. The region . . . is, of course, altogether valueless. Ours has been the first and will undoubtedly be the last, party of whites to visit the locality. It seems intended by nature that the Colorado River along the greater portion of its lonely and majestic way, shall be forever unvisited and undisturbed.

One may well wonder why Ives, a man intelligent enough to become an Army officer and to lead his party successfully over difficult terrain without loss, would have made such a silly statement. The answer: politics and money. At the time of the railroad siting explorations, of which Ives's was one, potential northern routes across the country vied with southern ones. Jefferson Davis, then director of the western surveys, favored a route across the 32nd parallel, which went through the heart of the American South. That would suit Southern interests better than the route along the 35th, which passed just south of the Grand Canyon. By damning the canyon country, Ives faithfully supported the claims of his mentor. In war, and even before it breaks out, truth is the first casualty.

Just before the Civil War began, Davis, by this time Secretary of War, placed Ives in charge of a survey of the southeastern boundary of California. The two decided to experiment to see how well camels would do in Death Valley, in the hope that these obdurate desert creatures might prove less so than the notoriously stubborn mules. But the camel experiment failed, as did Ives's next venture: serving as aide-de-camp to

Davis during the Civil War. Ives died in 1868 and is buried in Oxford, Mississippi. Though we remember him mainly for his seemingly absurd assessment of the potential of the Grand Canyon, Ives and his men were the first whites to reach the Colorado River from the rim.

Accompanying the Ives expedition was a young man named John Strong Newberry (1822–1892), an 1846 graduate of Western Reserve College and, two years later, of the Cleveland Medical School. In those days, before the Ph.D. degree had emigrated from Europe to America, a medical degree was the only post-graduate route open to one interested in science. Newberry spent the two years after medical school attending clinics and scientific lectures in Paris, an experience that seems to have lowered his interest in medicine and raised it in natural science. In 1855, he became assistant surgeon, geologist, and botanist to one of the Pacific Railroad surveys. Two years later, he joined Ives and traveled with his party down Diamond Creek to the Inner Gorge and the Colorado River. His excellent fieldwork allowed Newberry to construct the first stratigraphic column for the Grand Canyon.

In 1859, one year after the Ives expedition mustered out and ten years before the Major launched his boats at Green River, Newberry became chief scientist to a party led by Captain John N. Macomb (1811–1889) to explore the San Juan and the upper Colorado rivers. The report from this expedition did not appear until 1876, one year after Powell's exciting classic, and, as would become the case, Powell garnered all the attention. Showing no more apparent imagination than Ives, Macomb griped, "I cannot conceive of a more worthless and impracticable region than the one we now find ourselves in."

Having spent his time incarcerated in its canyons, Powell was literally not in the best position to draw conclusions about the overall geology of the Colorado Plateau. Thus, it is no surprise that in his 1875 report, he focused on broad geological principles, such as base level and the origin of valleys. In contrast, Newberry and the Macomb party traveled overland from Santa Fe all the way to the junction of the Green and Grand rivers. Working thousands of feet above where Powell would raft imprisoned, able to see one hundred miles, Newberry had the better perspective from which to understand regional geology. He was the first to see both the lower and upper Colorado River; his sound reconnaissance geology has held up.

Newberry quickly saw that the raw geology of the Colorado Plateau relegated European debates to the history of the science. It was obvious that the plateau rivers had carved their valleys and that all the erosion had taken place well above sea level. Perhaps with Ives's unfortunate summary judgment in mind, Newberry wrote that though "valueless to the agriculturalist, dreaded and shunned by the emigrant, the miner, and even the adventurous trapper, the Colorado Plateau is to the geologist a paradise. Nowhere on the earth's surface, so far as we know, are the secrets of its structure so revealed as here." In this geological land of Eden he had found "the most splendid exposure of stratified rock that there is in the world."

In his section of the Ives report, Newberry left no doubt as to whether ocean waves and currents, or rain and rivers, cause most erosion: "The peculiar topographical features" of the Plateau, the "great cañons of the Colorado, the broad valleys bounded by high and perpendicular walls *belong to a vast system of erosion, and are wholly due to the action of water* [Newberry's italics]. Probably nowhere in the world has the action of this agent produced results so surprising." Newberry's insight is all the more remarkable because he recognized that even in an area with almost no rainfall, running water is still the dominant geologic agent.

As to the origin of the great canyons and high cliffs, Newberry acknowledged the tempting answer that has occurred to so many visitors: "The first and most plausible explanation of the striking surface features of this region will be to refer them to that embodiment of resistless power—the sword that cuts so many geological knots—volcanic force. The Great Canyon of the Colorado will be considered a vast fissure or rent in the earth's crust, and the abrupt termination of the steps of the table-lands as marking lines of displacement." But, he wrote, "This theory though so plausible, and so entirely adequate to explain all the striking phenomena, lacks a single requisite to acceptance, and that is *truth* [Newberry's italics]. Aside from the slight local disturbance of the sedimentary rocks the strata of the tablelands are as entirely unbroken as when first deposited. I everywhere found evidence of the exclusive action of water in their formation." Citing the same evidence as Playfair, he wrote, "The opposite sides of the deepest chasm showed perfect correspondence of stratification, conforming to the general dip, and nowhere displacement."

But the question of erosion by running water "was only a subtext to the larger debate about the antiquity of the earth. A dramatic demonstration that fluvial [from running water] erosion had proceeded on an immense scale was an argument for a very old earth." In other words, a "vast system of erosion" requires a vast amount of time. Thus, Newberry, the very first geologist to see the canyon country, used its lessons to bury obsolete and limiting geological theories.

Newberry went on to one of the most illustrious careers of any nineteenth century American geologist, becoming director of the Ohio Geological Survey, professor at Columbia College, one of fifty founding members of the National Academy of Sciences, and president of both the American Association for the Advancement of Science and the New York Academy of Sciences. He became the lifelong enemy of his former student, Ferdinand Hayden, and a lifelong friend of John Wesley Powell.

Whatever Powell's faults, he did know how to pick friends and mentors. Throughout his career, he had them in high places—even the White House. In his post as the second director of the USGS, he hired geologists whom their peers would later judge to be the best that America has ever produced. Powell needed his own John Playfair, not because his scientific writing was opaque, but because it was so terse and usually lacking in evidence to support his conclusions. His assistants, as he called them and as they accepted with pride, would be the first to pass judgment on his theory of antecedence. Would they accept the conclusions of their leader, or would they politely request that he back them up with hard evidence?

PART TWO

Secrets of the River

Chapter 6

America's Greatest Geologists

I n 1869, a young man sought out Newberry to ask for a position with the Ohio Survey. Grove Karl Gilbert (1843–1918), age 26, from Rochester, New York, had already been to see the governor of Ohio, Rutherford B. Hayes, who told him that only Ohioans need apply. Newberry observed the same prerequisite, but, having spotted something special in the youngster, allowed him to join as a "volunteer" for 50 dollars per month to cover expenses. Gilbert had already had an unusual experience that would stand him in good stead throughout the rest of his geological career. For five years he had worked at Ward's Natural History Establishment in Rochester, "which provided and will still provide anything from trays of fossils to live black widows, from platypus eggs to relief maps, from laboratory insects to articulated skeletons of men or mastodons [one of which Gilbert helped excavate]." In its own way, especially during the Rochester winters, which tend to preclude field trips, the specimen-rich Ward's was a kind of geologist's paradise. Two years after Newberry took Gilbert on, Lieutenant George M. Wheeler (1842–1905), recruiting for his own survey west of the 100th Meridian, asked Newberry to recommend a geologist. He nominated Gilbert, who took west with him not only his mentor's endorsement, but also "Newberry's old maps, reports, and brilliant hunches."

The Wheeler Survey was one of four vying across the West for recognition, funding, and longevity. The leaders of the other three were Powell, Clarence King (1842–1901), and Ferdinand Hayden (1829–1887). In 1879, Congress consolidated the four into the United States Geological Survey, with King as its first director. In 1871, Wheeler and his party followed Ives's route, dragging their three boats upstream to the mouth of Diamond Creek. Thanks to the presence of Gilbert and geologist A. R. Marvine, the Wheeler expedition accomplished some excellent reconnaissance geology and mapping, but as usual it was Powell and his survey who garnered the attention of the public and of Congress. One of the reasons may have been Wheeler's turgid prose contrasted with Powell's exciting, first-hand account of his river adventure. Wheeler wrote that the canyons "stand without a known rival upon the face of the globe, and must always remain one of the wonders, and will, as circumstances of transportation permit, attract the denizens of all quarters of the world who in their travels delight to gaze upon the intricacies of nature." Compare that with Powell's vivid account of running the Canyon of Lodore—even his description of base level may be more exciting. Wheeler's decision to traverse Death Valley—in July—also reveals a near-fatal lack of timing. During the trip, one of the guides succumbed to heatstroke, requiring his companions to drape him over a mule; Gilbert's diary for the day recorded only "Breeze." Though Wheeler's report contained much useful information, it came too late to make a difference and was too ponderous to attract an audience.

In 1872, Gilbert visited the North Rim of the Grand Canyon and descended Kanab Creek to the river. During the trip he met members of the Powell party then exploring the region, including the Major's cousin Clement, the Major having left the second voyage by that point, and for the second time. Gilbert later met Powell in Salt Lake City; by 1875 he had accepted Powell's invitation to leave the Army-regulated Wheeler expedition and join his independent one. If anything more than Wheeler's prose were required to relegate the Wheeler Survey to oblivion, the defection of Gilbert did the job. He was irreplaceable. A century later, just before construction of the Glen Canyon Dam and the impoundment of Lake Powell would make the trip impossible, a group of geologists reenacted Powell's voyage. Sitting around their campfire one night, they fell to debating, as geologists will, who had been America's greatest. They chose Grove Karl Gilbert. Because Gilbert never took a professorship, had no students, wrote no textbooks, and had nothing but the excellence of his science to win him followers, his esteem is all the more remarkable.

By 1875, Powell's own report was out and his change of career from professor and field geologist to Washington man of influence was underway. In four years he would become the first Director of the new Bureau of Ethnology, having endorsed Clarence King to take the equivalent post at the newly consolidated U.S. Geological Survey. Like many of his undertakings, the tenure of the peripatetic King at the Survey was short-lived. Only two years after King began, he had resigned to seek his fortune in the Mexican gold fields. Powell succeeded him and until 1894 directed both the Survey and the Bureau of Ethnology. He continued in the latter post until his death in 1902.

Considering his distant base in Washington and his unusually onerous duties, Powell could no longer do field geology. He gradually lost interest in his maiden discipline, delegating the western fieldwork to the geologists of the fledgling Survey. In Gilbert and the next person in the story, Clarence Edward Dutton, Powell found two of the best ever. Gilbert was to measure the landscape of the Colorado Plateau with numbers, Dutton with figures of speech.

Gilbert took Powell's broad generalizations about base level and the evolution of rivers, systematized and quantified them, and tested them rigorously in the field. He "translated river silt and sand into equations [and] wrote a Euclidian treatise on erosion and structure." Where Powell appeared to be wrong, Gilbert took note but no pleasure, expressing his contrary opinion in the most diplomatic fashion possible. This was not sycophancy, but a mark of Gilbert's inherent kindness. One writer called him "the closest thing to a saint that American science has yet produced." Another, drawing on Gilbert's own phrase to describe the USGS, named him a "Great Engine of Research."

Gilbert's classic work, published in 1877, was entitled *The Geology of the Henry Mountains*, a small range located just northeast of Lake Powell and east of present-day Capitol Reef National Park. In the report, Gilbert explained that the Henry Mountains represented a new kind of geologic feature, a lozenge-shaped intrusion of igneous rock between sedimentary beds that had uplifted and domed the rocks above. He likened the force that had created the laccolith to a kind of subterranean hydraulic piston. Gilbert confirmed the conclusions of Newberry and Powell: the topography of the Colorado Plateau is due entirely to erosion by running water. But it was another section of his report that made Gilbert's reputation. In this primer on the work of stream erosion, transportation, and deposition, Gilbert enunciated geological principles that have held up since, ones that we will return to often.

With Gilbert, Powell may have made one of his rare mistakes. He restructured the Survey and brought Gilbert, one of the greatest field geologists who ever lived, back to Washington to ride, not one of Gilbert's beloved mules, but a desk. There Gilbert sat for twelve years. Had he spent those years in the field, surely they would have been at least as productive as his other fieldwork—and that is saying a great deal. Given what Gilbert accomplished in spite of his long sabbatical from the field, one can only imagine how much better off American geology would have been had he spent those dozen years west of the Rockies. What is harder to assess is how well off the U.S. Geological Survey would have been, during a critical period in its history, without Gilbert to guide it while his boss was off dealing with the politicians, at which he ultimately failed.

Powell's debt to Gilbert was enormous. Gilbert's report on the Henry Mountains appeared just when Powell was attempting to gain the support of Congress to fund the extension of his survey, lending it a much-needed credibility. But Gilbert had needed Powell to garner the Henry Mountains assignment in the first place. Geologist Bailey Willis knew them both. Willis was born four years before the Civil War and died four years after World War II; he wrote his first scientific paper in 1884 and his last in 1948. He and Gilbert were two of the first geologists hired by the new USGS. Willis said that Gilbert was "Powell's better half. I learned to know how much Gilbert, the true scientist, contributed to the geological thinking of Powell, the man of action. I do not think that they themselves were conscious of the degree to which the latter absorbed and gave out as his own ideas that the former had silently passed through." In his biographical tribute, Gilbert said that Powell:

Was extremely fertile in ideas, so fertile that it was quite impossible that he should personally develop them all, and realizing this he gave freely to his collaborators. The work which he inspired and to which he contributed the most important creative elements, I believe to be at least as important as that for which his name stands directly responsible. As he always drew about him the best ability he could command, his assistants were not mere elaborators, but made also important original contributions, and the ideas which he gave the world through others are thus so merged and mingled with theirs that they can never be separated.

Gilbert was tactful and respectful of Powell's seniority, but he was no toady. Powell actually had invented two classifications of structures and stream valleys, one of which Gilbert used and the other he avoided because, he said, it was not based on causes. Later in life, after his career in Washington had essentially ended, Powell turned to philosophy in a book entitled *Truth and Error*. He dedicated the book to his protégé, pioneer anthropologist Lester Ward. Earlier, Powell had praised a book by Ward, saying that it was "America's greatest contribution to scientific philosophy." Now it was Ward's turn to review Powell, but he failed to return the favor and panned the book.

In his eulogy, Gilbert allowed that Powell's philosophical ideas might have been "misunderstood," though Gilbert had to count himself as one of those who did so. Perhaps, Gilbert wrote diplomatically, Powell was ahead of his time, as he had shown himself to be in geology. Alas, in philosophy, it appears that Powell was not ahead of his time. No matter: Gilbert's valedictory to John Wesley Powell stands as a tribute to both:

To the nation he is known as an intrepid explorer, to a wide public as a conspicuous and cogent advocate of reform in the laws affecting the development of the arid West, to geologists as a pioneer in a new province of interpretation and the chief organizer of a great engine of research, to anthropologists as a leader in philosophic thought and the founder, in America, of the new regime.

A renowned historian of the mid-twentieth century, Bernard De Voto, summed up Powell's role in the history of the nation with a few simple words: "He was a great man and a prophet. Long ago he accomplished great things and now we are beginning to understand him." One of the things we now understand about Powell was his knack for spotting the most talented people and persuading them to join him.

AT AN EARLY AGE, POWELL and Gilbert had already started down paths that would lead them to geology. As his given name suggests, John Wesley Powell was the son of an itinerant Methodist preacher. He was born in 1834 in Mt. Morris, New York, but his restless father soon moved the family many hundreds of miles to a farm near Chillicothe, in southern Ohio. While still a lad there, young Wes came under the influence of one of those remarkable scholars who for reasons peculiar to their own history found themselves not in the academy but on the frontier. Wes's mentor, George Crookham, a giant of a man in every sense, had his own library, his own natural history museum, and an influential collection of friends, including the famous preacher Charles Grandison Finney, first president of Oberlin College. When the abolitionist views of Powell's father made things too hot for the boy at school, Crookham took him under his wing. Together they read the classics and went on field trips, sometimes accompanied by another of Crookham's friends, Ohio State University geologist William Mather, little Wes squeezed between the two men. Crookham, an accomplished "botanist, geologist, zoologist, ethnologist, archaeologist, historian, philosopher," belonged to the now-extinct species: "natural philosopher." As the youthful Wes grew to become the man John Wesley Powell, "his interests were by and large Crookham's interests." Without Crookham, John Wesley Powell would have amounted to something—though, it seems likely, something other than what he did.

Gilbert was born in Rochester, New York, nine years after Powell and was a quiet, sickly young fellow. His father, like Powell's, was unsuccessful; in Gilbert's case, a portrait painter barely able to provide for his family. Unlike Powell, Gilbert was fortunate enough to have a college in his hometown. Indeed, the itinerant Powell family had no hometown. Gilbert entered the University of Rochester at age fifteen, and in return for his impecunious family having scraped up the money for his tuition, agreed to get regular physical exercise. This family obligation may explain why Gilbert minored in geology, with its frequent field trips. He earned more credits in mathematics than in any other field, but also won an award for his Greek and was the president of two literary societies. After he graduated, Gilbert tried his hand at teaching public school in Michigan, but he found himself unable to discipline the rowdy schoolboys, barely younger than he was. Less than a year later, he returned to Rochester and Ward's.

Both Powell and Gilbert were good writers; at times the Major showed considerable flair. But no one would have mistaken either of them for artists. Powell's account of his danger-filled rides down the canyons, written in an exhilarating historical present, is a fine piece of adventure writing, but his scientific prose was so concise that only the inherent appeal of his subject carried it. Gilbert admitted that Powell "is terse to a fault. Usually he is satisfied with the simplest statement of his conclusions. Sometimes he adds illustrations. Only rarely does he explain them by setting forth their premises." Gilbert explained his own reasoning in such a logical way that it seemed impossible that no one had thought of his findings before, as though "he were simply reporting the relative weights from a scale." One incident shows how Gilbert's logical mind would not let him avoid addressing even two such intractable subjects as government bureaucracy and recalcitrant livestock. On assignment in Salt Lake City, he called attention "to the fact that there were a variety of brands on the government mules," and went on to "respectfully suggest the propriety of the adoption by the Director of a uniform pattern."

For a different reason in each case, the writings of Powell and Gilbert have stood the test of time. But to expect the two of them to have come up with prose to match the sublimity of the Grand Canyon would be to ask too much—that would prove beyond the powers of all but one.

Clarence Edward Dutton (1841–1912) came to geology much later in life than Powell and Gilbert and arrived through the back door of happenstance. He attended Yale, graduating in the same class as the famed fossil collector Othniel C. Marsh. As if to demonstrate what scientific training can permit, Marsh had stepped off the Union Pacific train in Nebraska for a few hours and discovered Eohippus, thus helping to solidify the gathering support for Darwin's theory of evolution. At Yale, Dutton, self-described as "omnibiblical," was a good enough writer to win the Yale literary prize. He began training for the ministry at the Yale Divinity School, but lost his faith and left "before he was thrown out," winding up studying chemistry and mathematics instead. In 1862, Dutton joined the 21st Connecticut Volunteers, serving as First Lieutenant, Adjutant, and Captain respectively and seeing action at Fredericksburg, Nashville, and Petersburg. He took a competitive examination to transfer from the Volunteers to the Ordnance Corps and became one of only three to pass.

At the end of the war, Dutton found himself stationed at the Watervliet Arsenal in Troy, New York, with time on his hands. His curious mind and the proximity of the Bessemer Steel Works led him to study the chemistry of molten metals and the technology of iron and steel production, inspiring a lifelong interest in volcanoes and igneous rocks. Another important institution was nearby: the Albany Paleontological Museum, where Dutton learned geology from James Hall, mentor to several eminent geologists. In 1870, he transferred to the Frankfort Arsenal and a year later to Washington, D.C., where he "insinuated" himself into the scientific societies of the capital and "cultivated the Survey men and became well acquainted with Powell." The Major also knew how to cultivate important people, including Ulysses S. Grant, under whom he had served at Shiloh. Using that connection, Powell and Joseph Henry, Secretary of the Smithsonian Institution, persuaded the War Department to detail Dutton to the Powell Survey in 1875. He was to remain with it and its successor, the USGS, for the next fifteen years. Dutton was also a good friend of Clarence King and shared his tastes. But while King was openly devout, Dutton was not. Possibly in part for this reason, Dutton may have felt more comfortable with the Powell's Survey than King's. After all, it was Powell who refused to stop his river expeditions to observe the Sabbath.

Even though, in Dutton's words, no one was as well suited to describe the geology of the Grand Canyon as Powell, "whose right to do so [was] virtually prescriptive," Powell's increasing responsibilities in Washington meant that he had to delegate the work on the Colorado Plateau to others. Powell needed geologists to match the country; in Gilbert and Dutton he found them. Powell understood that he needed to delegate completely, without what today we call, but Dutton never would have, micro-management. Gilbert's tribute to Powell's willingness to give away his best ideas we have seen. Dutton wrote, "[Powell's] direction of the Survey has not been limited to the perfunctory duties of an administrative officer. On the contrary he has furnished those who he has called to his assistance with methods of observation and principles which have worked like a master-key in opening to our understanding the meaning of this wonderful region." Elsewhere he acknowledged that he could not tell where his own ideas began and those of his colleagues ended, saying: "if a full accounting were called for it would bring me to bankruptcy."

If Gilbert was Powell's right hand, Dutton was the left. Both were better field geologists than Powell; he was surely a better politician than they. Powell paid Dutton the finest compliment possible: he entrusted him with the task of exploring and reporting on the geology of the Plateau Country that Powell loved so much. Though it seems clear that Powell intended to anoint Dutton his geological successor, Dutton's broader interests took him far from the Grand Canyon. In the end, politics came between them and they parted bitterly.

Given the generous credit that the three were ready to grant each other, it seems safe to say that Powell would have been the first to agree that in the Heavens of Geology, one could have no better "assistants" than Grove Karl Gilbert and Clarence Edward Dutton. Powell put the Big Cañon on the map; Gilbert forced it to bend to logic; Dutton rendered it Grand.

Chapter 7

The Sublimest Thing

~~~~~~

Hurtling through the depths of the dark canyons, never knowing what lay around the next bend, low on rations and crew morale, John Wesley Powell was not in the best position or frame of mind to comprehend the overall geology of the Colorado Plateau. Near the end of his first trip, the canyons had become "our granite prison." Powell regretted that "all around me are interesting geological records. The book is open, and I can read as I run. All about me are grand views, for the clouds are playing again in the gorges but somehow I think of the nine days' rations, and the bad river, and the lesson of the rocks and the glory of the scene is but half seen." He faced the same task as a military officer under fire: how to capture his objective while getting his troops out alive. What had become important about rocks was avoiding those in the riverbed, not contemplating those exposed in the canyon walls. Understandably, Powell began to draw his major geological insights on the Green, a country where canyons have a known end, voyageurs have barely sampled their rations, and time is available for rest and reconnaissance.

Gilbert had visited the Grand Canyon several times and written a fine treatise on the lessons it offers. But his assignments were mainly elsewhere: first in the Henry Mountains and then back in Washington, far from his beloved western field. Had Powell assigned Gilbert to work out the geological history of the Grand Canyon district, he would have done his usual superb job. People today would still read a Gilbert report on the Grand Canyon for its geology, though not for its prose. Instead, Powell gave the task to the one geologist who could understand that the Colorado Plateau country "should be described in blank verse and illustrated on canvas"; one who could "forget that he is a geologist and feel himself a poet."

Of Clarence Edward Dutton's three books on the geology of the Colorado Plateau, the one that has brought him lasting fame is *The Tertiary History of the Grand Cañon District*. As an educated man of the world, Dutton knew that the country to which Powell had "called" him was both a scenic and geologic innovation, one whose proper description required attention to both art and science. His report was itself an innovation, a combination of travelogue and astute geological analysis. Its unusual approach required Dutton to acknowledge at the outset that the monograph "departed from the severe ascetic style which has become conventional in scientific manuscripts." But "no apology [was] called for," because "under ordinary circumstances the ascetic discipline is necessary. Give the imagination an inch and it is apt to take an ell, and the fundamental requirement of scientific method—accuracy of statement—is imperiled. But in the Grand Canyon district there is no such danger. The stimulants which are demoralizing elsewhere are necessary here to exalt the mind sufficiently to comprehend the sublimity of the subjects." No one could ever accuse Dutton of conventionality. The Grand Canyon, the most unconventional of geologic scenes, provided exactly the right stimulus for his mind and his pen.

The result was "a rich and embroidered nineteenth century traveler's prose [that] flows around bastions of geological fact as some of the lava coulées on the Uinkaret flow around gables of sedimentary strata"; "an extraordinary ensemble of the science, aesthetics, cartography, painting, photography, illustration, and ideas that have animated the intellectual and imperial expansion of America." Dutton's style is all the more remarkable when we learn that he memorized long passages and dictated them, pacing the room, omnipresent cigar in hand. The modern author, pecking away at a word processor, salivates to learn that in one such session, Dutton dictated 11,000 words and then needed to make only two small corrections. His facility at dictation may account for the conversational tone of his writings—and there is independent evidence that Dutton could hold his own with the most adept conversationalists, Clarence King and Henry Adams included, in a day when conversation mattered.

As he had recommended, Dutton approached the Grand Canyon as an artist contemplates a blank canvas:

> The lover of nature, whose perceptions had been trained in the Alps, in Italy, Germany, or New England, in the Appalachians or Cordilleras, in Scotland, or Colorado, would enter this strange region with a shock, and dwell there for a time with a sense of oppression, and perhaps with horror. Whatsoever things he had learned to regard as beautiful and noble he would seldom or never see, and whatsoever he might see would appear to him as anything but beautiful and noble. Whatsoever might be bold and striking would at first seem only grotesque. The colors would be the very ones he had learned to shun as tawdry and bizarre. The tones and shades, modest and tender, subdued yet rich, in which his fancy had always taken special delight, would be the ones which are conspicuously absent. But time would bring a gradual change. Some day he would suddenly become conscious that outlines which at first seemed harsh and trivial have grace and meaning; that forms which seemed grotesque are full of dignity; that magnitudes which had added enormity to coarseness have become replete with strength and even majesty; that colors which had been esteemed unrefined, immodest, and glaring, are as expressive, tender, changeful, and capacious of effects as any others. Great innovations, whether in art or literature, in science or in nature, seldom take the world by storm. They must be understood before

they can be estimated, and must be cultivated before they can be understood.

Any science editor would likely have rejected this passage as overly ornate. But Dutton had to satisfy only his mentor, editor, and publisher, Major Powell, who, one suspects, could hardly have been more pleased with Dutton's report, having had such literary aspirations himself.

It is a rare piece of scientific writing that bears reading more than a century later for any reason other than historical research, though Gilbert did pen several such rarities. Typically, both prose style and scientific conclusions have long since gone out of date. Indeed, new facts and interpretations have superseded Dutton's ideas about the history of the Grand Canyon. But because of its art, *Tertiary History* has survived not only Dutton but also several eras in the history of geology. Likely it is timeless.

Dutton differed from Powell and Gilbert in an important way: his mind tended more to ultimate causes. The Colorado Plateau may have risen like a log into a saw, but what lifted the log? If large blocks of the earth's surface can ascend by thousands of feet, or tens of thousands, can they also descend by similar amounts? And does the rising and sinking of so much rock not require that the interior of the earth be able to flow like a liquid? Because warmer rocks are more fluid, does this not in turn mean that the earth has a warm, internal zone near its surface? Could this zone be the source of the volcanic rocks that, since his days near Bessemer, had interested Dutton? And what then was the source of the extra heat required to warm some parts of the earth's interior more than others? During Dutton's day, geologists could only ask such questions; they had far too little collateral information to be able to answer them. It is a tribute to Dutton's mind that he was able to go beyond the field work and literary triumph that surely would have more than satisfied anyone else to identify some of the outstanding problems of geology.

GEOLOGY OF THE HIGH PLATEAUS OF UTAH was Dutton's first western monograph. The title referred to the lofty uplands of the central and southeastern sections of the state, well to the north and west of the Grand Canyon, that bear such exotic names as Páhvant, Tushar, Markágunt, Sevier, Paunságunt, and Aquarius. We might as well call them mountains, for the Tushar reaches above 12,000 feet, almost twice the elevation of Mt. Mitchell, the highest peak east of the Mississippi, and the other peaks of the High Plateaus often exceed 11,000 feet. Yet their tops tend to be relatively flat, so we deem them plateaus.

When in 1874 Powell urged him to prepare a report on the geology of the High Plateaus, Dutton declined, writing later that he "distrust[ed] my fitness for the work." But a year later, when Powell again proffered the job, Dutton accepted. Though not known for a lack of confidence or false modesty, even Dutton appears to have been daunted by the country he was asked to explain, noting that the resulting work contained "many imperfections and . . . falls far short of my hopes and expectations." He admitted that he lacked experience in this new sort of geology, but probably "the magnitude of the task proposed was too great even for experienced observers." Of course, with the exception of Gilbert, who escorted Dutton to the High Plateaus, and Powell, no one had more experience. The High Plateaus, like the Colorado Plateau, are a geological innovation for which no other terrain can have prepared one.

Utah, in Stegner's felicitous phrase, "has a spine like a Stegosaurus." If the dinosaur faced north, the Wasatch Range would make up the section from the middle to the head. The High Plateaus would reach from the mid-section toward the tail, stretching south-southwest from around Nephi, Utah, on modern Interstate 15, for 175 miles and extending east-west for 25 to 80 miles. In Figure 1, they occupy the high ground stretching northeast from the Grand Canyon nearly to the Uintas. Zion Canyon is carved into the southwestern edge of the southernmost of the Plateaus, the Markágunt; Bryce Canyon is carved into the southeastern edge of the Paunságunt. The Aquarius, farthest to the northeast, looks east to Gilbert's Henry Mountains and Capital Reef

National Park, and south to Grand Staircase-Escalante National Monument, 6,000 feet below. One hundred or so miles to the south lie Glen and Grand Canyons.

The Plateaus stand so high because they are capped with resistant volcanic rocks that held up while erosion carried out a great denudation that stripped away thousands of feet of softer rock all around:

And now the relation of the High Plateaus to the Plateau Province at large becomes evident. They are the remnants of great masses of Tertiary and Cretaceous strata left by the immense denudation of the Plateau Province to the south and east. From the central part of the province the Tertiary beds have been wholly removed and nearly all of the Upper Cretaceous. A few remnants of the Lower Cretaceous stretch far out into the desert, and one long narrow causeway, the Kaiparowits Plateau, extends from the southeastern angle of the district of the High Plateaus far into the Central Province and almost joins the great Cretaceous mesas of Northeastern Arizona, being severed from them only by the Glen Cañon of the Colorado. The Jurassic has also been enormously eroded. The average denudation of the Plateau Province since the closing periods of the local Eocene can be approximately estimated, and cannot fall much below 6,000 feet.

From one vantage point atop the Plateaus and looking to the south and east, perhaps at a moment when the poet was in a certain mood, he saw, "A picture of desolation and decay; of a land dead and rotten, with dissolution apparent all over its face. It consists of a series of terraces, all inclining upwards to the east, cut by a labyrinth of deep narrow gorges, and sprinkled with numberless buttes of strange form and sculpture."

But from a spot above Zion Canyon, the view inspired:

From the southwest salient of the Markágunt we behold one of those sublime spectacles which characterize the loftiest standpoints of the Plateau Province. Even to the mere tourist there are few panoramas so broad and grand; but to the geologist there comes with all the visible grandeur a deep significance. The radius of vision is from 80 to 100 miles. We stand upon the great cliff of Tertiary beds which meanders to the eastward till lost in the distance, sculptured into strange and even startling forms, and lit up with colors so rich and glowing that they awaken enthusiasm in the most apathetic. To the southward the profile of the country drops down by a succession of terraces formed by lower and lower formations which come to the daylight as those which overlie them are successively terminated in lines of cliffs, each formation rising gently to the southward to recover a portion of the lost altitude until it is cut off by its own escarpment. Thirty miles away the last descent falls upon the Carboniferous, which slowly rises with an unbroken slope to the brink of the Grand Cañon. But the great abyss is not discernible, for the curvature of the earth hides it from sight. Standing among evergreens, knee-deep in succulent grass and a wealth of Alpine blossoms, fanned by chill, moist breezes, we look over terraces decked with towers and temples and gashed with cañons to the desert which stretches away beyond the southern horizon, blank, lifeless, and glowing with torrid heat. To the south-westward the Basin Ranges toss up their angry waves in characteristic confusion, sierra behind sierra, till the hazy distance hides them as with a veil. Due south Mount Trumbull is well in view, with its throng of black basaltic cones looking down into the Grand Cañon. To the southeast the Kaibab rears its noble palisade and smooth crest line, stretching southward until it dips below the horizon more than a hundred miles away. In the terraces which occupy the middle ground and foreground of the picture we recognize the characteristic work of erosion. Numberless masses of rock, carved in the strangest fashion out of the Jurassic and Triassic strata, start up from the terraced platforms. The great cliffs—perhaps the grandest of all the features in this region of grandeur—are turned away from us, and only now and then are seen in profile in the flank of some salient. Among the most marvelous things to be found in these terraces are the cañons; such cañons as exist nowhere else even in the Plateau Country. Right beneath us are the springs of the Rio Virgin, whose filaments have cut narrow clefts, rather than cañons, into the sandstones of the Jura and Trias more than 2,000 feet deep; and as the streamlets sank their narrow beds they oscillated from side to side, so that now bulges of the walls project over the clefts and shut out the sky. They are by far the narrowest chasms, in proportion to their depth, of which I have any knowledge.

By the final chapter in *High Plateaus*, Dutton is at his best:

The Aquarius should be described in blank verse and illustrated upon canvas. The explorer who sits upon the brink of its parapet looking off into the southern and eastern haze, who skirts its lava-cap or clambers up and down its vast ravines, who builds his camp-fire by the borders of its snow-fed lakes or stretches himself beneath its giant pines and spruces, forgets that he is a geologist and feels himself a poet. From numberless lofty standpoints we have seen it afar off, its long, straight crest-line stretched across the sky like the threshold of another world. We have drawn nearer and nearer to it, and seen its mellow blue change day by day to dark somber gray, and its dull, expressionless ramparts grow upward into walls of majestic proportions and sublime import. The form-less undulations of its slopes have changed to gigantic spurs sweeping slowly down into the painted desert and parted by impenetra-ble ravines. The mottling of light and shadow upon its middle zones is resolved into groves of *Pinus ponderosa*, and the dark hues at the summit into myriads of spikes, which we know are the storm-loving spruces.

The ascent leads us among rugged hills, almost mountainous in size, strewn with black bowlders, along precipitous ledges, and by the sides of cañons. Long detours must be made to escape the chasms and to avoid the taluses of fallen blocks; deep ravines must be crossed, projecting crags doubled, and lofty battlements scaled before the summit is reached. When the broad platform is gained the story of "Jack and the beanstalk," the finding of a strange and beautiful country somewhere up in the region of the clouds, no longer seems incongruous. Yesterday we were toiling over a burning soil, where nothing grows save the ashy-colored sage, the prickly pear, and a few cedars that writhe and contort their stunted limbs under a scorching sun. To-day we are among forests of rare beauty and luxuriance; the air is moist and cool, the grasses are green and rank, and hosts of flowers deck the turf like the hues of a Persian carpet. The forest opens in wide parks and winding avenues, which the fancy can easily people with fays and woodland nymphs. On either side the sylvan walls look impenetrable, and for the most part so thickly is the ground strewn with fallen trees, that any attempt to enter is as serious a matter as forcing an abattis. The tall spruces (*Abies subalpina*) stand so close together, that even if the dead-wood were not there a passage would be almost impossible. Their slender trunks, as straight as lances, reach upward a hundred feet, ending in barbed points, and the contours of the foliage are as symmetrical and uniform as if every tree had been clipped for a lordly garden. They are too prim and monotonous for a high type of beauty; but not so the Engelmann spruces and great mountain firs (*A. Engelmanni, A. grandis*), which are delightfully varied, graceful in form, and rich in foliage. Rarely are these species found in such luxuriance and so variable in habit. In places where they are much exposed to the keen blasts of this altitude they do not grow into tall, majestic spires, but cower into the form of large bushes, with their branchlets thatched tightly together like a great hay-rick. Upon the broad summit are numerous lakes—not the little morainal pools, but broad sheets of water a mile or two in length. Their basins were formed by glaciers, and since the ice-cap which once covered the whole plateau has disappeared they continue to fill with water from the melting snows.

Dutton concludes *High Plateaus* figuratively seated atop the south-western edge of the Aquarius, "at a high pass that is the main divide between the drainage systems of the Colorado and the Great Basin." Here the lava-cap of the Aquarius ends and, protruding from underneath it, the Tertiary beds thrust out a long, rocky peninsula to the south, at an altitude of nearly 11,000 feet. From this lofty perch Dutton could see a grand spectacle, a "great number of cliffs rising successively one above and beyond another, like a stairway for the Titans, leading up to a mighty temple." And even farther in the distance, "the platform of the Kaiparowits Plateau . . . a member of the Kaibab system . . . a broad causeway, reaching to the Colorado [and] Glen Canyon." He had come to the end of the *High Plateaus*; ahead lay *The Tertiary History of the Grand Cañon District.*

TERTIARY HISTORY WAS THE FIRST monograph printed by the new USGS. The Survey intended the book for specialists and printed only 3,000 copies, causing Stegner to note a paradox: the more copies of a book printed, the fewer that remain in circulation decades later. "Print something in an edition as large as that of a Sears Roebuck catalogue, and people do not value it; they wrap eggs in its pages, start fires with it, hang it in the backhouse. But print an edition of fifty numbered copies, and a hundred years later forty-eight of them will survive, whereas of the Sears Roebuck catalogue there remains only one torn example." Stegner knew whereof he spoke, for he wrote his Ph.D. dissertation on Dutton. But as a "penniless graduate student during the Depression," he could not afford a copy of *Tertiary History*. In those days before copying machines, to have a duplicate that he could mark up, Stegner remembered that he "was reduced to typing the entire book. But as soon as I could afford it," he wrote, "I paid the price, because it was a book I wanted to own."

The book's appeal came not only from Dutton's happy marriage of science and art, but from its illustrations. Accompanying Dutton were three men, each of whom was in his own discipline as much an artist as Dutton. Their work added a priceless dimension to his geology and word paintings. The photographs of Jack Hillers (1843–1925), cameraman on Powell's second expedition, appeared in the book as heliotypes. He had become, in Stegner's opinion, the best photographer of Indians. Both Thomas Moran (1837–1926), the most famous painter of western scenes, and William Henry Holmes (1846–1933), geologist and artist, were along.

Congress paid $10,000 for Moran's massive "Chasm of the Colorado" and hung it in the Capitol opposite his "Grand Canyon of the Yellowstone." Thirty years after Moran first saw the Grand Canyon (of the Colorado), he wrote, "Of all places on earth the great canyon of Arizona is the most inspiring in its pictorial possibilities." In each of the last twenty-five years of his life, Moran returned to his inspiration to paint hundreds of different views.

Stegner considers Holmes's pen-and-ink drawing, "Panorama from Point Sublime," to be "the most magnificent picture of the Grand Canyon ever drawn, painted or photographed." The atlas of *Tertiary History* included a larger, color version, from which I have taken the cover of this book. Not only does Holmes capture the majesty of the Grand Canyon, the two figures he draws in the foreground have intrigued every student of the canyon. One man is seated on a rock precipice, paper and pad in hand. The second man leans over, his hand on the other's shoulder in friendship, to admire his work. Logic would suggest that Dutton, who dictated his reports during the winters back in Washington rather than composing them seated on the canyon rim, is admiring the seated Holmes. But to leave the identity of the figures a mystery does no harm. Just the opposite: now the geologist can admire the artist's rare gift of capturing the vista on paper; the artist can admire the ability of the geologist to envision how it all came to be and the vast amount of time required. As Worster put it, "The Grand Canyon challenges both the artist and scientist to redefine their relationship to one another and, together, to reeducate their perception of the earth."

Holmes was to prove himself even more versatile than Powell or Dutton. After leaving the Survey in 1884, he returned to ethnology and in 1902 succeeded Powell as chief of the Bureau of American Ethnology. Late in life he directed the National Gallery of Art.

High Plateaus ended with Dutton gazing from atop the Aquarius down and across his "stairway of the Titans" to the distant canyon country below. *Tertiary History* opens from another of the High Plateaus, looking south:

Before the observer who stands upon a southern salient of the Markágunt is spread out a magnificent spectacle. The altitude is nearly 11,000 feet above the sea, and the radius of vision reaches to the southward nearly a hundred miles. In the extreme distance is the calm of the desert platform, its surface mottled with indistinct lights and shades, too remote to disclose their meaning. Against the southeastern horizon is projected the pale blue escarpment of the Kaibab, which stretches away to the south until the curvature of the earth carries it out of sight. To the southward rise in merest outline, and devoid of all visible details, the dark mass of Mount Trumbull and the waving cones of the Uinkaret. Between these and the Kaibab the limit of the prospect is a horizontal line, like that which separates the sea from the sky. To the southwestward are the sierras of the Basin Province, and quite near to us there rises a short but quite lofty range of veritable mountains, contrasting powerfully with the flat crestlines and mesas which lie to the south and east.

Dutton begins by describing the great denudation that has removed thousands of feet of sedimentary rock; he ends at the Great Unconformity exposed deep in the Grand Canyon, where Paleozoic rocks overlie ancient Precambrian ones (see Figure 16, page 260). Today we know from radiometric age dating that the gap in time between the two represents more than one billion years. In *Tertiary History*, a chapter of geology alternates with one of descriptive prose, in which Dutton invites the reader to ride alongside him on three imaginary journeys. The literary strategy of bringing the reader along on an imaginary trek may have been a natural for the western geologist, because Ferdinand Hayden used it before Dutton. "I shall ask the reader to travel with me along the line of the Union Pacific Railroad," Hayden wrote in 1870, "to study rocks and unearth their fossil contents; and in many a locality we shall find the poet's utterance no fiction. . . ."

The first of Dutton's three rides began at Kanab, Utah, and proceeded west to Zion Canyon and the valley of the Virgin River, illustrated by Holmes's magnificent painting, "Smithsonian Butte," the frontispiece of *Tertiary History*. Had copies of Dutton's description of this trip been available to the public, they would surely have inspired countless young men to forget becoming a scout, cowboy, outlaw, rancher, rustler, mountain man, sheep herder, or any other western profession than geologist:

Late in the autumn of 1880 I rode along the base of the Vermilion Cliffs from Kanab to the Virgin, having the esteemed companionship of Mr. Holmes. We had spent the summer and most of the autumn among the cones of the Uinkaret, in the dreamy parks and forests of the Kaibab, and in the solitudes of the intervening desert; and our sensibilities have been somewhat overtasked by the scenery of the Grand Canyon. It seemed to us that all grandeur in beauty thereafter beheld must be mentally projected against the recollection of those scenes, and be dwarfed into a commonplace by the comparison; but as we moved onward the walls increased in altitude, in animation, and in power. At length the towers of Short Creek burst into view, and, beyond, the great cliff in long perspective thrusting out into the desert plain its gables and spurs. The day was a rare one for this region. The mild, subtropical autumn was over, and just giving place to the first approaches of winter. A sullen storm had been gathering from the southwest, and the first rain for many months was falling, mingled with snow. Heavy clouds rolled up against the battlements, spreading their fleeces over turret and crest, and sending down curling flecks of light mist into the nooks and recesses between towers and buttresses. The next day was rarer still, with sunshine and storm battling for the mastery. Rolling masses of cumuli rose up into the blue to incomprehensible heights, their flanks and summits gleaming with sunlight, their nether surfaces above the desert as flat as a ceiling, and showing, not the dull neutral gray of the east, but a rosy tinge caught from the reflected red of rocks and soil. As they drifted rapidly against the great barrier, the currents from below flung upward to the summits, rolled the vaporous masses into vast whorls, wrapping them around the towers and crest-lines, and scattering torn shreds of mist along the rock-faces. As the day wore on the sunshine gained the advantage. From overhead the cloud-masses stubbornly withdrew, leaving a few broken ranks to maintain a feeble resistance. But far to the Northwest, over the Colob, they rallied their black forces for a more desperate struggle, and entered with defiant flashes of lightning the incessant pour of sun-shafts.

For the second trip, Dutton moves upriver to the Toroweap overlook on the North Rim, some ninety miles west of where the Grand Canyon National Park would later have its headquarters on the South Rim. Here he could look down on the river and around him to great faults and to the basaltic rocks that always fascinated him. Finally, and from the length and enthusiasm of his description, happily, he approaches the Kaibab Plateau, looming above the eastern end of the Grand Canyon, almost as if unaware of what awaits him:

We continue to cross hills and valleys, then follow a low swale shaded by giant pines with trunks three to four feet in thickness. The banks are a parterre of flowers. On yonder hillside, beneath one of these kingly trees, is a spot which seems to glow with an unwonted wealth of floral beauty. It is scarcely a hundred yards distant; let us pluck a bouquet from it. We ride up to the slope. The earth suddenly sinks at our feet to illimitable depths. In an instant, in the twinkling of an eye, the awful scene is before us. Reaching the extreme verge the packs are cast off, and sitting up on the edge we contemplate the most sublime and awe-inspiring spectacle in the world. We named it *Point Sublime* [Dutton's italics].

Such scenes "must be cultivated before they can be understood." Only then is one ready to recognize with Dutton that the Grand Canyon is "the sublimest thing on earth . . . not alone by virtue of its magnitude, but by virtue of the whole—its ensemble." To a man of Dutton's time, aesthetic sensibility, and training, the word sublime was among the most powerful in the language—the last he would use lightly. One should not construe the sublime as merely beautiful or moving; rather, as Donald Worster expresses it, "The experience of the sublime lay beyond reason; it defied human understanding, control, or activity. To feel the sublime was to be swept by incomprehensible powers." Thus, to encounter sublimity was not merely to feel pleasure; it was to be awestruck and diminished by powers beyond one's intellectual grasp, powers of frightening implications. To stand at the brink of the Grand Canyon is to come face to face with the "blind forces of nature." It is to "feel like mere insects crawling along the street of a city flanked with immense temples, or as Lemuel Gulliver might have felt in revisiting the capital of Brobdingnag, and finding it deserted."

Dutton had shown that as an artist he had cultivated and understood the Grand Canyon better than anyone. Indeed, the canyon does tempt the geologist to forget that calling and consider himself a poet, though most would be wise to resist the temptation. Now it is time to draw the artist down to earth from poetic altitudes. Dutton worked not for a literary house that could indulge his tastes, but for the hard-headed U.S. Geological Survey. He had to demonstrate that he comprehended, and could explain, the geological history of the Colorado River and the Grand Canyon. In doing so, he would inevitably have to pass judgment on his mentor's theory of antecedence. He did both, and went further. His work in the Grand Canyon District led him to discover the roots of a revolution in geology that would take a century to arrive.

Chapter 8

Earth's Engine

Dutton's predecessors—Newberry, Powell, and Gilbert—had worked out the stratigraphic column of the Colorado Plateau and identified the great faults and other prominent geologic features. Dutton did not have to start from square one, but could meld their work and his own observations into a geological history of the Grand Canyon.

Key to any geologist's interpretation is the ability to use the features of sedimentary rocks—fossils, minerals, cross-beds, mud cracks, and ripple marks—to determine the environment in which the parental sediments accumulated. Fossils show whether a rock accumulated in salt water or fresh, and if in salt water, at what depth. Cross-bedded sandstones indicate deposition as sand dunes, shale deposition in a swamp, and so forth. Finding in the Cretaceous rocks evidence of marine deposition, Dutton concluded that a sea had occupied the region at that time. The rocks of the next-younger period, the Tertiary, showed evidence of deposition partly in a shallow sea and partly above sea level, indicating that as the land rose, the Cretaceous sea shrank and the water turned brackish (part salt and part fresh).

Rocks of the Eocene Epoch are widespread across the Colorado Plateau. They are exposed from the Uinta Mountains on the north, to the Wasatch on the west, and well south into Arizona. These thick deposits accumulated in freshwater. Thus, by the Eocene—some 50

Era	Period	Epoch	Millions of years ago
CENOZOIC	QUATERNARY	Holocene	0.01
		Pleistocene	1.8
	TERTIARY	Pliocene	5.3
		Miocene	23.8
		Oligocene	33.7
		Eocene	54.8
		Paleocene	65
MESOZOIC	CRETACEOUS		144
	JURASSIC		206
	TRIASSIC		248
PALEOZOIC	PERMIAN		290
	CARBONIFEROUS		354
	DEVONIAN		417
	SILURIAN		443
	ORDOVICIAN		490
	CAMBRIAN		543
PRECAMBRIAN	PROTEROZOIC		2,500
	ARCHEAN		

Figure 5.
Geologic timescale.

million years ago—the Cretaceous Sea had disappeared entirely, to be replaced by an enormous lake, one that Dutton thought might have been as large as all of today's Great Lakes combined. The deepest sections of the lake lay over present-day Marble and Grand Canyons. From the way successively younger rocks changed their character, Dutton could tell that "the lake shrank away very slowly towards the north, finally disappearing at the base of the Uintas at the close of Eocene time."

Dutton inferred that after the lake dried up, "upon the floor of this basin, as it emerged, a drainage system was laid out. Such a drainage system would necessarily conform to the slopes of the country then existing. The configuration of the principal channels would be very much like that at present. The trunk channel would flow southwestward and westward, while the tributaries would enter it on either hand very much as the larger and older tributaries now do." Thus, Dutton believed that the courses of the rivers of the Colorado Plateau were determined by the locations at which they happened to find themselves in the wake of the departed Eocene lake. The rivers must themselves date back to the Eocene and be the oldest features around. Dutton agreed with Powell: continuing uplift of the Colorado Plateau meant that, in order to maintain their courses, the rivers had been required to cut vertically instead of laterally. The rivers were there before the structures; in Powell's classification, they are antecedent.

Dutton believed that the ancestral Colorado River, its course as fixed by its history as ours by our genes, had flowed south from the Uinta Basin, across the Colorado Plateau, and on past the mouth of the Virgin River. For comparison, Dutton asked his reader to imagine that the Great Lakes region were to rise by 2,000 feet, a fraction of the uplift of the Colorado Plateau. Then:

In no great length of time Ontario would be drained by the St. Lawrence, lowering its channel, and that river would become one with the Niagara. The same process would be repeated at Erie, Huron, and Superior, the lakes vanishing and leaving only a great river with many branches. Such was the origin of the Colorado: first a Hellespont, then a St. Lawrence, then a common but large river heading in the interior of a continent.

The geologists of Dutton's day, hoping to have their necessarily descriptive science rank beside quantitative ones such as physics, sought to identify the principles that govern the earth and to express them as laws. Gilbert was especially prone to doing so. Emulating him, Dutton went beyond Powell's simple declaration that "the river had the right of way" to enunciate "The Law of Persistence of Rivers." Without it, a geologist in the Plateau Country "would see little but Sphinxes"; with the law, the geologist could "translate many mysteries." Dutton said the Law held because:

Of all the changing features of a continent the least changeful are its great rivers. Undoubtedly rivers have perished and undoubtedly they have shifted parts of their course somewhat; but on the whole their tenacity of life is wonderful, and the obstinacy with which they sometimes maintain their positions is in powerful contrast with the instability of other topographical features.

But this explication of the law amounts to a mere assertion. Indeed, Dutton offered little scientific evidence to back up the newly defined law. Even so, Dutton said, nowhere is the law more applicable than in the Colorado Plateau:

It would be difficult to point out an instance where a great river has ever existed under conditions more favourable to longevity and stability of position than those of the Colorado and its principal tributaries. Since the epoch when it commenced to flow it has been situated in a rising area. Its springs and rills have been among the mountains and its slope has throughout its career been continuously though slightly increasing. The relations of its tributaries have in this respect been the same, and indeed the river and its tributaries have been a system and not merely an aggregate, the latter dependent upon and perfectly responsive to the physical conditions of the former. And now we come to the point. The Colorado and its tributaries run today just where they ran in the Eocene Period. Since that time mountains have risen across their tracks, whose present summits mark less than half their total uplifts; the river has cleft them down to their foundations. The Green River, passing the Pacific Railway, enters the Uintas by the Flaming Gorge, and after reaching the heart of this chain, turns eastward parallel to its axis for thirty miles, and then southward, cutting its way out by the splendid canyon of Lodore. Then following westward along the southern base of the range for five miles, a strange caprice seizes it. Not satisfied with the terrible gash it has inflicted upon this noble chain, it darts at it viciously once more, and entering it, cuts a great horse-shoe canyon more than 2,700 feet deep, and then emerging, goes on its way. Thenceforward, through a tortuous course of more than 300 miles down stream the strata slowly rise—the river almost constantly running against the gentle dip of the beds, cutting through one after another, until its channel is sunk deep in the carboniferous. Further down, near the head of Marble Canyon, the Kaibab rose up to contest its passage, and a chasm of more than 6,200 feet in depth bears witness to the result. It is needless to multiply instances. The entire province is a vast category of instances of drainage channels running counter to the structural slopes of the country. I am unable to recall a single tributary to the right bank of the Colorado which does not somewhere, and generally throughout the greater part of its course, run against the dips.

And on to the key that unlocks:

The structural deformations of the region—the faults, flexures, and swells, had nothing to do with determining the present distribution of the drainage. The rivers are where they are in spite of them. What then determined the situation of the present drainage channels? The answer is that they were determined by the configuration of the old Eocene lake-bottom at the time the lake was drained. Soon after the surface began to be deformed by unequal displacement, but the rivers had fastened themselves to their places and refused to be diverted. This, then, is the key which unlocks for the geologist the vestibule of the Plateau Country. The rivers were born within the country itself, they are older than its cliffs and canyons, older than its great erosion—the oldest things in its Tertiary history; nay, they are its history, which we may yet read imperfectly in their canyon walls. The mountains and plateaus are of subsequent origin. They arose athwart the streams only to be cleft asunder to give passage to the waters. The rivers amid all changes have ever successfully maintained their right of way.

Dutton took antecedence almost to a *reductio ad absurdum*, but, like Powell, failed to present much in the way of geological evidence. Rather, Dutton simply declared the entire Colorado River to be older than each structure and obstacle it crosses, even saying that "it is needless to multiply instances." He did not appear to consider Powell's other option for a formerly consequent stream, superposition, which could also produce incised canyons.

Dutton surely knew that he would be far from the last geologist to enjoy Newberry's paradise. His successors surely would make countless new observations and uncover countless new facts; indeed, they still uncover them today. As a man of science and the world, Dutton knew that the fate of most scientific theories is to wind up altered, if not entirely replaced. Why men of the intelligence of Powell and Dutton did not either provide more compelling evidence for their conclusions, or present them with a modicum of caution, remains a mystery. But they were far from the only scientists to state their theories with unwarranted certainty.

To one studying the history of science, it even seems that the more certain are the proponents of a theory, the more likely they are to be wrong, though like politicians, when the truth outs, they are usually no longer around. In Powell and Dutton's day, Lord Kelvin pronounced, from a height so lofty that only a physicist could attain it, that the earth is as rigid as steel and could not possibly be older than twenty-four million years. Kelvin also declared that a heavier-than-air machine could never fly. The long-lived Bailey Willis (1857–1949) pronounced that "the great ocean basins are permanent features of the earth's surface and they have existed, *where they are now* [italics added], with moderate changes of outline, since the waters first gathered." One of Willis's last papers, written in 1944, he entitled "Continental Drift, ein Märchen (a fairy tale)." In the 1960s, British scientist Sir Harold Jeffreys, and others, condemned continental drift as not merely doubtful, but physically impossible. As the space age dawned, some students of the moon believed that volcanic activity created every single feature on its surface and said so with confidence—only to have meteorite impact turn out to be the cause of most of those features. To go on would be instructive, but depressing. Each of us, especially scientists, would do well to keep in mind Goethe's admonition: "To be uncertain is to be uncomfortable but to be certain is to be ridiculous."

Before we leave the subject of certainty in science, the reader will have noted that so far in this account, G. K. Gilbert and the USGS have remained blameless. Alas, just as we can justly criticize Powell for failing to mention the brave men who risked their lives for him, and Dutton for going too far in promulgating as a Law what was no more than a possibility, so can we take even the saintly Gilbert to the geological woodshed. The object of his error lay only a few score miles to the southeast of the Grand Canyon, where jagged fragments of an extraterrestrial iron meteorite lay scattered around a large hole in the ground. People called the site Coon Mountain for the upraised rim of the depression; later it became known as Meteor Crater.

In 1891, Gilbert heard a talk describing a "circular elevation . . . occupied by a cavity" near the Painted Desert of Northern Arizona. He had thought enough about the craters on the moon to recognize that impacting asteroids might have created both the lunar craters and the Arizona one, thus explaining the presence of the iron fragments around the latter. The crater was not volcanic, for no lava was in evidence. But there was one other possible terrestrial explanation: steam might have exploded from underground, blasting out the cavity but leaving no telltale igneous rock behind. Numerous examples of such steam explosions occur around the world, though meteorite fragments do not surround the depressions that they leave.

Gilbert, who prided himself on his logicality and his ability to use the scientific method, came up with ways to test whether a meteorite impact or a steam explosion created Meteor Crater. First, if an underground steam explosion had blasted out the cavity, geologists would find, lying around on the surface nearby, the rock that once occupied the hole. The volume of the excavated rock should be the same as the volume of the crater. On the other hand, if a "star entered the hole the hole was partly filled thereby and the remaining hollow must be less in volume than the rim." Second, if an iron meteorite had created the crater, the floor ought to conceal the "buried star," whose magnetism would betray its presence.

Before conducting his field tests, Gilbert gave 800 to 1 odds that they would favor the meteorite impact option; otherwise, how to explain the meteorite fragments? But both tests failed to confirm his prediction. The volume of material ejected from the crater turned out to be the same as the volume of the crater itself (by coincidence, we now know) and there was no anomalous magnetism (because impacting meteorites blast themselves to smithereens, we now know).

Having set the rules, Gilbert was not the sort to change them after the game was over. Meteorite impact having failed the tests, he had to conclude that steam volcanism and not impact had created the crater, even though that required him to attribute the meteorite fragments to coincidence and come down on the wrong side of exceedingly long odds. Any of us can understand how Gilbert could hoist himself atop such a petard of treacherous logic. What is harder to understand is that for the remaining two decades of his life, Gilbert refused to discuss Meteor Crater or even to acknowledge new findings that went against his conclusion. So did the USGS: rather than run the risk of contradicting the Magister, the Survey put research on meteorite impact off limits. "The long shadow that Gilbert cast over impact studies stretched to 1959, when another Survey geologist falsified Gilbert's conclusion about Meteor Crater. In part for his work in showing that impact created Meteor Crater, and with no small irony, in 1983 Eugene Shoemaker (1928–1997) won the G. K. Gilbert Award of the Geological Society of America."

⌐⌐

DUTTON WAS RIGHT ABOUT far more than he was wrong. Had the opinionated Bailey Willis paid more attention to one line of Dutton's reasoning, he might have avoided, or at least toned down, his dogmatic, long-lived opposition to the notion that continents can move—though on second thought, he probably would not have. Dutton's views about the history of the Colorado Plateau have required revision, but the fundamental observations he made there led him to insights of even greater importance. One can trace a direct line from Dutton's thinking, inspired by what he saw in the Grand Canyon and the Colorado Plateau, to the theory of continental drift and from there to the plate tectonic revolution in geology. Plate tectonics is so important that following this trail is worth a brief side trip from our main story.

Searching as he always did for ultimate causes, Dutton came to ponder how erosion and deposition (sedimentation) might relate to the rise and fall of the earth's surface. Which was cause and which was effect, he wondered. He understood that erosion and deposition are intimately related:

Erosion viewed in one way is the supplement of the process by which strata are accumulated. The materials which constitute the stratified rocks were derived from the degradation of the land. This proposition is fundamental in geology—nay, it is the broadest and most comprehensive proposition with which that science deals. It is to geology what the law of gravitation is to astronomy. Erosion and "sedimentation" are the two half phases of one cycle of causation—the debit and credit sides of one system of transactions. The quantity of material which the agents of erosion deal with is in the long run exactly the same as the quantity dealt with by the agencies of deposition; or, rather, the materials thus spoken of are one and the same. If, then, we would know how great have been the quantities of material removed in any given geological age from the land by erosion, we have only to estimate the mass of the strata deposited in that age. Constrained by this reasoning the mind has no escape from the conclusion that the effects of erosion have been vast. If these operations have achieved such results, our wonder is transferred to the immensity of the periods of time required to accomplish them; for the processes are so slow that the span of a life-time seems too small to render those results directly visible.

Dutton's insight began with the geologic column of the Plateau, where as noted he and his predecessors had found a sequence of marine (deposited in sea water) sedimentary rock thousands of feet thick. Similarly thick marine sequences occur in the Appalachians, the Alps, and other mountain ranges. The simplest way for a thick mass of sediment to accumulate would seem to be for a deep oceanic trough to receive deposits over a long period of time and gradually fill up. In that case, the oldest rocks in the sequence, those that settled first, on the very bottom of the trough, would show evidence of having accumulated in deep water. The youngest, at the top, would have formed almost at sea level. But this reasonable prediction fails: even in the thickest of typical sedimentary sequences, the evidence shows that each rock layer accumulated in shallow water. For the marine rocks of the Colorado Plateau, this evidence consists of pieces of petrified wood, shallow-water ripple marks, cross-bedding indicative of sand dunes, coal and peaty layers, shallow-water marine fossils, and the like. There is no doubt that these rocks formed in shallow water.

Dutton described the problem this way: "Here we are confronted by [a] paradox . . . a tract which is rising yet sinking; a basin which is shallow, which receives a great thickness of deposits, and yet is never full." But such a paradox must be apparent and arise only from the human brain, not from Nature herself, where things happened as they did, whether or not we can explain them. The solution to the apparent paradox, Dutton realized, is that as each sedimentary layer accumulates—in shallow water—it weighs down just enough to cause the earth beneath it to sink and make way for the next layer, which then also deposits in shallow water, and so on. The result is a thick set of shallow-water beds: not a paradox, but the product of a fluid earth.

But, Dutton wondered, if deposition and subsidence have a cause-and-effect relationship, might not erosion and uplift have the same?

Those areas which have been uplifted most have been most denuded. I have asked myself a hundred times whether we might not turn this statement round, and say that those regions which have suffered the greatest amount of denudation have been elevated most, thereby assuming the removal of the strata as a cause and the uplifting as the effect; whether the removal of such a mighty load as ten thousand feet of strata from an area of ten thousand square miles may not have disturbed the earth's equilibrium of figure, and that the earth, behaving as a *quasi*-plastic body, has reasserted equilibrium by making good a great part of the loss by drawing upon its whole mass beneath.

The notion that erosion could induce uplift, rather than the other way around, made sense to Dutton, for "few geologists question that great masses of sedimentary deposits displace the earth beneath them and subside. Surely the inverse aspect of such a problem is *a priori* equally palpable." In other words, if the weight of a great mass of sediment can cause a block at the earth's surface to subside, could not the removal of a similar mass cause such a block to rise? If the earth's surface can rise and sink by thousands of feet, somehow at depth, over geologic time, it must be able to flow.

Assuming that the weight of sediments offers a plausible cause of subsidence, what initiates uplift? Dutton recognized that without the answer, "the explanation is not quite complete." The best he could do was to refer to "that mysterious plutonic force which seems to have been always at work, and whose operations constitute the darkest and most momentous problem of dynamical geology." This clandestine agent must "at least inaugurate, and perhaps in part . . . perpetuate, the upward movement, but . . . we must also recognize the cooperation of that tendency which indubitably exists within the earth to maintain the statical equilibrium of its levels." This question "turns on the rigidity of the earth." When he viewed the Plateau Country through his experienced eyes, Dutton found it "easy to believe that . . . the rigidity of the earth may become a vanishing quantity."

The key to the mystery had already turned up, halfway around the world, in a serendipitous discovery in India. There, in the 1850s, Surveyor-General Colonel George Everest found a small but significant difference between the latitudes of two stations depending on whether he measured them with a surveyor's chain or using astronomical observations. A simple explanation presented itself: the gravitational attraction of the nearby Himalaya Mountains had pulled the surveyor's plumb bob (a conical metal weight dangled at the end of a cord) off true vertical. Everest referred the problem to John Pratt, the Archdeacon of Calcutta and a mathematician. Pratt determined that the mountains had not deflected the bob as much as they should have; indeed, near the coast, the bob actually swung away from the mountains. Because a plumb bob responds only to the mass of objects, the Himalayas, which stand so high, must be made of rock less dense than average, and that less-dense rock must extend down well beneath sea level. In other words, a mountain must be like an iceberg: some of its mass rises up where we can see it; most hides beneath the surface. Icebergs float in a denser medium; somehow mountains must do the same. Like Archimedes's wooden block floating in water, mountains, plains, and ocean basins must float in at least rough gravitational balance according to their dimensions and density. Pratt and George Airy, the Astronomer Royal, debated exactly how the earth accomplished this balancing act, but both models required that the earth's interior be able to flow.

───

DUTTON WROTE SEVERAL ARTICLES on the evidence for a fluid interior, and in 1889, seven years after *Tertiary History*, delivered his most complete statement in a paper presented to the Philosophical Society of Washington, entitled "On Some of the Greater Problems of Physical Geology." He began by posing the fundamental questions of his science: "The greatest problems of physical geology I esteem to be: 1st, What is the potential cause of volcanic action? 2d, What is the cause of the elevation and subsidence of restricted areas of the earth's surface? 3d, What is the cause of the foldings, distortions, and fractures of the strata?" Each process requires prodigious amounts of energy. In effect, Dutton asked: What is the engine that drives the earth? That would remain the key, unanswered question until the 1960s.

The first problem was without solution: theories regarding the source of volcanism go "to pieces at the touch of criticism." The second, the cause of uplift and subsidence, was simply an enigma. But a solution to the third problem looked "more hopeful." Before explaining the reasons for that hopefulness, Dutton had to address the then-reigning model of earth behavior: the contraction theory. Geologists of the day, including John Wesley Powell, believed that the earth's interior was cooling and therefore that the earth was contracting. This followed from the nebular theory for the origin of the solar system and the earth, which held that the sun and earth had initially been hot and had been cooling ever since, the incorrect assumption that had led Kelvin into error. As the interior of the earth cools and shrinks away from the brittle surface above, large blocks were supposed to have broken out and settled down to become ocean basins, leaving the regions in between to stand high and become continents.

Contraction requires that the forces that act on surface blocks be mainly vertical. That did seem to be the case in the Colorado Plateau, but, as Dutton noted, in the Appalachians, the Alps, and other mountain ranges, the folding forces had come from the side. They acted horizontally, not vertically. Furthermore, the Appalachians comprise a central belt of folded rocks with an ocean floor on one side and a 2,000-mile-wide, undeformed plain on the other. Contraction should instead have produced more or less uniform wrinkling around the globe, like the shriveled apple skin that provided an analogy. Dutton wrote, "In short, [contraction] could not form long, narrow belts of parallel folds. I dismiss the hypothesis with the remark that it is quantitatively insufficient and qualitatively inapplicable." Before long, everyone had come to agree with him and the contraction theory was extinct. Unfortunately, this left geologists without a comprehensive theory of earth behavior, one that would explain why the earth has continents and ocean basins, and what happens within them. They would not have such a theory confirmed until the 1960s.

Dutton next noted that the earth is not a perfect sphere, but bulges at the equator, exactly as would a fluid body its size, spinning on its axis. This correspondence constitutes *prima facie* evidence that overall, the earth does behave as a fluid. That the earth's less-dense continents stand high while its more-dense ocean basins ride low is now understandable: each is merely obeying its own density, subject to the same force, gravity, that produces the earth's overall figure. Where the material at the surface is lighter, it bulges; where it is denser, it flattens or sinks. Dutton said, "For this condition of equilibrium of figure, to which gravitation tends to reduce a planetary body, irrespective of whether it be homogeneous or not, I propose the name Isostasy [Dutton's italics; from the Greek for equal standing]."

Dutton cited several pieces of evidence for isostasy. One is the just-discussed observation that each part of a thick sequence of sedimentary rocks formed in shallow water, revealing how a trough of deposition sank under the weight of the accumulating sediments. Another came from the opposite geological setting: mountain platforms like the High Plateaus of Utah that are both highly eroded and elevated. Like finding a thick wedge of shallow-water sediments, at first this combination seems paradoxical: erosion lowers elevations; therefore, the more erosion, the lower a region ought to be. Hearkening back to one of his original questions, Dutton concluded that deeply eroded, but lofty, mountains showed that as erosion removed material, the decreased weight allowed the block to rise, subjecting it to still more erosion and establishing the connection between erosion as cause and uplift as effect. A third type of evidence came from measurement of the pull of gravity at different points on the earth's surface. In Dutton's day, such measurements were primitive, but scientists had made enough to confirm that mountains and other elevated regions, like the Himalayas, indeed had lower densities than the ocean basins and depressed areas.

Dutton pointed out that as the land surface erodes, streams transport the resulting sediments to the continental margins, where they accumulate; their weight then causes the interior to flow out of the way to make room for them. But where does the departing internal material go? He thought it moved sideways, toward the interior of the continent, where it intrudes and uplifts the surface, causing more erosion, transportation, and deposition. Thus is set up a cycle, or feedback loop. To those who say that rock cannot flow, Dutton pointed to the glaciers of Greenland, whose ice, though brittle in the short run, obviously flows in the long run.

Again, where most would have been satisfied to take the subject this far, Dutton went further. He closed his paper by noting that because erosion lowers the high places and deposition fills in the low, the earth's surface should approach a uniform level. Because it does not, some independent force more powerful than isostasy—that is, more powerful than gravity—lifts the surface by thousands of feet in one place and pulls it down in another. This force must not be the same as the one that folded the rock layers of mountain ranges such as the Appalachians, because some folded regions have not been uplifted, while some uplifted blocks, such as the Colorado Plateau, are not folded. And in many places—the Sierras, for example—the folding took place in one geologic age and the uplift in a much later one. The best Dutton could do, again, was to identify the problem: "The cause which elevates the land involves an expansion of the underlying magmas, and the cause which depresses it is a shrinkage of the magmas. The nature of the process is, at present, a complete mystery." But even as he wrote, experiments were under way in European laboratories that would lead to a revolution in physics and solve this geological mystery of mysteries.

The revolution began in 1895, three years after Dutton's paper appeared, when Roentgen detected X-rays. His discovery led a few months later to Becquerel's discovery of radioactivity, a term coined by his student, Marie Curie, who went on to receive the Nobel Prize in both physics and chemistry. Marie's husband, Pierre, found that the element radium not only maintains a temperature greater than its surroundings, but that one gram emits 100 calories per hour—enough to raise the temperature of a gram of water from the freezing to the boiling point. Because common rocks contain radium, uranium, and thorium—radioactive elements all—scientists suddenly understood that the earth has an abundant but previously unrecognized internal source of heat. Lord Kelvin had assumed that the earth had only the heat left over from its formation and had been cooling ever since; radioactivity invalidated his assumptions and his (and Clarence King's) conclusion that the earth is only 24 million years old. As historian Naomi Oreskes put it, "atomic energy could power not only the world but also geological theory."

Dutton soon understood that radioactivity could solve the first of his three great problems: the potential cause of volcanic action. He knew from his geological work that "the mean rigidity of the subterranean masses is inferred to be far less than that of ordinary surface rocks." If the earth were rigid, isostasy would be impossible, but geologic evidence had established isostasy as a fact. Further, getting now to the great problem, somehow the earth melts internally and the melts reach the surface, where they flow out as lava. But what would cause a largely solid earth to warm up enough to first become fluid and then to melt? Dutton postulated that the answer is the heat released by radioactive decay.

Most geologists of the early twentieth century accepted Dutton's notion that isostasy allows large blocks of the earth to move up and down in a fluid layer. If so, could they not also move sideways? Geologic orthodoxy, as expressed by Bailey Willis's statement that the continents have always been "where they are now," said no, they could not. But one who was not a geologist, German meteorologist and polar explorer Alfred Wegener, disagreed and went on to develop the idea of drifting continents into a full-fledged theory, which he published in 1915. Like Dutton, Wegener rejected contraction. He said it violated not only isostasy, but the pet theory of paleontologists at the time: the notion that long fingers of continental rock had once bridged the continents. Paleontologists required these land bridges because, had the land masses always been in the same place, there was no way to explain the presence of identical fossils on continents now separated by thousands of miles of sea water. Paleontologists had either to accept the anathema of continental drift or conjure up a raft of serpentine isthmuses that connected continents and provided a dry route for migrating organisms. Conveniently, when no longer needed, the land bridges had sunk back beneath the waves, out of sight but not out of mind. But are land bridges consistent with isostasy? To stand above sea level and thus allow the sunlight and sustenance that biota require, land bridges would have to have been made of less-dense, continental rock. But according to isostasy, less-dense rock does not sink—it floats. Wegener reached the conclusion that everyone would reach sometime over the next half-century: land bridges violated isostasy and were as mythical as Atlantis.

But, Wegener wrote, "If . . . continental blocks really do float on a fluid, there is clearly no reason why their movement should only occur vertically and not also horizontally, provided only that there are forces in existence which tend to displace continents, and that these forces last for geological epochs." In other words, if blocks of the earth can float up and down in a fluid layer, one cannot logically reject out of hand the notion that these same blocks, or even whole continents, can move sideways. All it would take is a horizontal force sufficiently strong to push them. But few gave credence to Wegener's seemingly irrefutable logic.

One reason they did not is that whereas the force behind isostasy—gravity—is well known, no one knew of any force that could propel continents sideways. Thus, according to the self-serving lore of geology, Wegener's theory failed "for lack of a mechanism." Ironically, British geologist Arthur Holmes had proposed an acceptable mechanism—a convecting earth mantle—in the late 1920s, but geologists failed even to consider it. They should have retained continental drift as a working hypothesis while they looked for the mechanism, but because drift appeared to violate their most central tenet—the uniformitarianism of Hutton and Lyell—they proclaimed such a search unnecessary, even heretical. In 1967, some fifty years after Wegener, the twin discoveries of paleomagnetism and seafloor spreading led geologists to reverse themselves and accept continental drift, in its new cloak of plate tectonics, driven by the mantle convection that Arthur Holmes had clearly described four decades earlier.

It is worth a brief aside to compare the two symbiotic sciences of biology and geology. It has been said that "evolution, for the first time, gave biologists something to do." Before Darwin's theory appeared in the mid-nineteenth century, biology was necessarily descriptive, without an overarching theory. Geologists believed that they did have such a theory: contraction. But after it failed at the turn of the century, and geologists rejected continental drift in the 1920s, they were left with no model of earth behavior. There was no explanatory framework on which to hang even the most obvious geologic observations. One result of this lack from the 1950s can serve as an example. By that decade, it was clear that forces operating from the southeast had compressed the rocks of the Appalachian Mountains. Yet to the southeast of the continental United States lies only the floor of the Atlantic Ocean. Even the best geologists, when asked the origin of these forces, where now only blue water resides, could only shrug their shoulders. Geology remained largely descriptive for a century after biology, but the power of plate tectonics and the invention of absolute age dating and other powerful techniques produced a burst of creative energy for geology. Indeed, one can make a good case that the last decades of the twentieth century have been geology's golden age.

But if the mantle could not behave as a fluid, the Colorado Plateau could not have risen, continents could not have drifted, and plates could not have moved. Thus, from Dutton's search for the cause and meaning of uplift and subsidence in the Colorado Plateau, one can draw a straight line to his papers on isostasy and radioactivity. From there only a short step takes one to Wegener's insistence that continents that bob up and down can also slip sideways, then on to Holmes's theory of convection, and thence to the eventual acceptance of continental drift and plate tectonics. Dutton, riding from the hamlet of Kanab to the buttes of the Virgin, crossing the basalt coulees of the Uinkaret to Toroweap Overlook, ambling his beloved Kaibab, plucking bouquets, only to find that the earth suddenly fell away beneath his feet to leave him standing on the edge of a great precipice, could not have known that he also stood on the edge of a revolution in geology. Yet, perhaps the urbane poet-geologist would not have been surprised to learn that most of a century would pass before the revolution took hold. Dutton knew that "great innovations, whether in art or literature, in science or in nature, seldom take the world by storm. They must be understood before they can be estimated, and must be cultivated before they can be understood."

By the time Powell and Dutton had finished their work on the Grand Canyon, they believed they had settled the question of the origin of that "great innovation." Dutton had not only embraced Powell's theory of antecedence, he had taken it to an extreme. According to the inviolable Law of Persistence of Rivers, the entire Colorado River, from north of the Uintas to the Gulf of California, is older than each structure it crosses. But Mother Nature and fellow geologists often resist the simple explanation. The unpoetic but logical Gilbert had seen enough to know that the story was more complicated than his two colleagues had decided. He would require evidence for any claim, no matter who made it. For antecedence to prevail, the theory would first have to get by Gilbert. If it had trouble doing so, the door would be open for other critics.

Chapter 9

Where Everything
Is Exposed

~~~~~~

G ilbert was a better field geologist than either Powell or Dutton.
His logical mind left him disinclined to accept assertions at
face value. The combination led him to disagree not with
deceased predecessors, but with his boss and his colleague, his two closest professional associates. But Gilbert handled the task with his usual
tact. Only one year after the Major's *Exploration* appeared, at the end of
an article entitled "The Colorado Plateau Province as a Field for Geological Study," he gently expressed a different opinion as to the origin of the
drainage systems on the Colorado Plateau. But first, Gilbert set out the
many advantages the region holds for the geologist. They stand as a
testament to why the Grand Canyon has drawn generations of earth
scientists. First is the climate:

> The air is so dry that there is no turf, no accumulation of humus,
> often no soil, and so little vegetation that the view is not obstructed.
> From a commanding eminence one may see spread before him, like
> a chart, to be read almost without effort, the structure of many
> miles of country, and in a brief space of time may reach conclusions,
> which, in a humid region, would reward only protracted and labori-
> ous observation and patient generalization. There is no need to
> search for exposures where everything is exposed.

The second advantage is that the rivers of the Plateau remove the rocks from the uplands "as fast as they decay, and soil cannot accumulate. Thus does a thorough drainage conspire with aridity to prepare for the geologist a land of naked rock." Third are the canyons, which "bear the same relation to the plain that geological cross sections do to a geological map. They introduce in all categories of observation a third dimension, and enable the contemplation of all the phenomena of structure with reference to depth as well as length and breadth." The final advantage came from what was not present: the glacial moraines that blanket so much of the bedrock of the upper Midwest and Northeast. But, Gilbert had to admit, the Plateau did have one drawback: volcanic lavas cover one-tenth of its sedimentary rock and do an even better job of obscuring bedrock than glacial drift. Ironically, when scientists discovered how to use radioactivity to measure rock ages, the lava turned out to be key to understanding the modern history of the Colorado River in the Grand Canyon.

Another advantage of the Colorado River, to the eternal gratitude of Powell and his men, is that in spite of cutting through mountain ranges and across every kind of geologic structure, and in spite of presenting the rafter with well over 100 white-water rapids, the river has not a single waterfall. Gilbert wondered why. As he often did, in coming up with the reason he hit upon a crucial geologic principle, though one that would take others several decades to appreciate. The absence of waterfalls even led Gilbert to an important insight into the origin of the Grand Canyon itself. Here is how he began his explanation:

Where rivers descend a slope that is terraced by the alternation of hard and soft strata, they are apt to leap from the edges of the hard beds in waterfalls. But the Colorado, not withstanding the structure of its bed presents the most favorable conditions, makes no leap. At the head of Marble Canyon, for instance, the river crosses a great bed of limestone, lying nearly level and underlaid by a great bed of friable sandstone. The limestone resists all erosive agents as strongly as does the Niagara limestone, and the sandstone yields to them as easily as does the Niagara shale. But, instead of plunging from one to the other in a great cataract, the Colorado cuts the two with nearly equal grade of channel.

If the Colorado and Niagara rivers do not behave differently because they flow over different rock types, the difference must be due to "some condition that pertains to the constitution of the stream itself." That condition, Gilbert wrote, "is to be found in the relation of corrasion [abrasion] to transportation." In other words, the two rivers differ in the balance between the amount of sediment that erosion provides to them and their ability to carry that sediment away.

Gilbert asked his reader to imagine a stream with a constant supply of water. After flowing for a while, the stream will have picked up and be transporting as much sediment as it can. As long as "its velocity remains the same, it will neither corrade [abrade mechanically] nor deposit, but will leave the grade of its bed unchanged." But if the slope over which the stream travels decreases, so will its velocity, and "its capacity for transportation will become less than the load, and part of the load will be deposited." If the slope were instead to increase, "the capacity for transportation will become greater than the load, and there will be corrasion of the bed." Gilbert understood that in this way, a stream constantly adjusts so as to balance each of the factors that affect it. "A stream, which has a supply of debris equal to its capacity, tends to build up the gentler slopes of its bed and cut away the steeper. It tends to establish a single, uniform grade." According to Stephen Pyne, Gilbert, writing when railroads loomed large in our nation's consciousness, took the concept and name of a graded stream "from the engineering term used to describe the longitudinal profile of railroad tracks."

From this reasoning, Gilbert drew an important lesson: "The differentiation will proceed until the capacity for corrasion is everywhere proportioned to the resistance, and no farther—that is, until there is an equilibrium of action. In general, we may say that a stream tends to equalize its work in all parts of its course." Geologists called Gilbert's concept dynamic equilibrium. Streams, like other natural systems, adjust the factors that affect them so as to do minimal work. We will meet this idea again later.

To explain the Colorado River's lack of waterfalls, Gilbert, in a Goldilocks analogy, used the notion of dynamic equilibrium and the contrast provided by the Niagara and Platte Rivers. Lake Erie provides the Niagara with its water and traps sediment that the river would otherwise carry. This provides clear water for Niagara Falls downstream from Lake Erie; the sediment-free water erodes its channel rapidly so as to regain its equilibrium load. In this part of the Great Lakes region, a layer of hard limestone overlies a soft shale, setting up the ideal condition for a waterfall to form and retreat upstream. The Niagara River plunges over the ledge of the resistant limestone into a splash pool below and strikes the weak shale, which quickly erodes until the now unsupported limestone cliff above collapses, moving the waterfall upstream and starting the process anew. In the 12,000 years since the last glacier left to open the way, Niagara Falls has retreated about seven miles, three feet per year. At that rate, in only an eyeblink of geologic time, Niagara Falls must disappear.

In contrast to the Niagara, sediment so chokes the Platte that the river can barely move, much less erode. The average gradient of the Platte is as steep as that of the Colorado, but its soft banks, and the sand and soil blown in from the dusty plains that surround it, caused sediment to clog the Platte's channel. Thus, the Platte cannot lower its bed any faster than erosion lowers the general level of the surrounding plains. As a result, the Platte has neither rapids nor waterfalls.

Like the "just right" bed selected by Goldilocks, the Colorado evidently lies midway between the Niagara and the Platte, having just the right amount of suspended load to allow it to erode any obstacle in its bed. "Where the bedrock is soft, the declivity is small. Where it is hard, the declivity is relatively great; but in the toughest hornblende rock the mean angle of slope does not exceed three degrees." In other words, as fast as an incipient waterfall raises its head, the Colorado River lops it off. But how then to explain the many rapids in the Colorado River? Boulders that flash floods have washed down the side canyons and into the main stem, rather than bedrock protuberances, cause these rapids. Each large flash flood in the Grand Canyon adds new boulders to the main stem, altering existing rapids or creating new ones and keeping life interesting for river runners.

The absence of waterfalls in the Grand Canyon—or indeed anywhere on the Colorado—is more than a happy condition for river runners. The striking, counterintuitive lack of waterfalls in such a deep and long canyon must reveal something special about the history and evolution of the Colorado River. What does it mean that not only has the main-stem Colorado cut down so rapidly as to incise a deep canyon, it has eroded so effectively as to obliterate every obstacle in its path, including any incipient waterfalls and bedrock rapids? It means that some geologic factor, or combination of factors, has given the Colorado River the erosive strength of ten. One key to answering the question, "What caused the Grand Canyon?" is to identify that factor or factors.

As exemplified in his most visible mistake, Meteor Crater, Gilbert required himself to apply observation and rigorous logic to geologic problems. At the end of his article on the many advantages of the Colorado Plateau, it came time to face, as he called it, "The Problem of Inconsequent Drainage." The phrasing revealed that Gilbert did not accept Powell's conclusion that everywhere along the Colorado River's path, it is older than the structures it crosses. Instead, a "problem" existed, one that by implication Powell had not solved after all. Though Dutton's *Tertiary History* would not appear for several years, the three men were in close contact, and Gilbert must have known that Dutton would transmogrify the aphorism of their mentor into a full-fledged Law, and thus that Gilbert would be on record as disagreeing with Dutton as well.

Before making his doubt clear, Gilbert elaborated on Powell's classification of drainage systems. Consequent streams, as the word indicates, owe their paths to the underlying geologic structure and therefore flow from the crests of anticlinal hills down to the troughs of synclinal valleys. If erosion goes on for any length of time, a consequent stream must shift to some other type. If the land over which a stream flows begins to rise, Gilbert wrote, "Unless the displacements are produced with unusual rapidity the waters will not be diverted from their accustomed ways. The effect of local elevation will be to stimulate local corrasion, and each river that crosses an uplifted block will, inch by inch as the block rises, deepen its channel and valorously maintain its original course. It will result that the directions of the drainage lines will be independent of the displacements." This is antecedence, but Gilbert gave no sign that he regarded it as so dominant as to prove the putative "law of persistence of rivers."

Gilbert's careful wording reveals a fundamental problem with the idea of antecedence. What if "the displacements *are* [italics added] produced with unusual rapidity" and the river cannot cut down rapidly enough to keep up? Because the water must go somewhere, the river then must divert to another course. Powell understood that had the Uintas risen rapidly enough, the down-cutting Green River would not have been able to keep pace, and instead of carving the Canyon of Lodore, would have relocated somewhere else. Indeed, we have no way of knowing that the Green did not try out several other courses before settling into the present one. Antecedence reveals only the last path taken.

As a reminder, Powell's third possibility, superposition, comes about when sedimentary layers accumulate on top of, and eventually bury, an older geologic structure like a mountain range. Later, consequent streams begin to erode the uppermost layer of the sedimentary cover, unaware, so to speak, of the resistant rocks that lie hidden in wait below. As the streams gradually lower the overall landscape, one eventually impinges on the buried structure. If it can, the river then maintains its course by cutting down more rapidly and vertically than in adjacent sections of the river, where it may still be running over softer sedimentary rock. If the river cannot cut down fast enough, it too must divert to some other course.

An antecedent river could cut a gorge like the Canyon of Lodore, but so could a superposed one. Telling them apart requires far more than a cursory examination. An antecedent stream requires uplift on a broad scale; a superposed one requires sedimentary burial and erosion on an equally broad one. Because these are regional events, all the rivers in a given area are apt to have the same origin: the Green River is not likely to be antecedent while its tributaries, such as the nearby Yampa and White Rivers, are superposed, and vice versa. This means that in order to understand the origin of a river, the geologist needs to know a great deal about the general geology of an area, far more than Powell was able to learn in his brief exposure, much of it spent deep in the canyons. Indeed, it was not until the 1980s, well over a century after Powell declared the Green antecedent, that its origin finally appeared settled.

Gilbert's failure to accept at face value the judgment of Powell and Dutton that the Green and all the Plateau rivers are antecedent gave a hint of the difficulty the theory would face. But in his polite way, Gilbert avoided openly disagreeing with his mentor and his colleague:

> A large share of the drainage of the Plateaus is not consequent. How much is super-imposed, and how much antecedent remains to be determined. With the solution of the problem are involved the determination of the antiquity and history of the Green and Colorado Rivers, and the physical history of the great Tertiary lakes; and we may hope that from its discussion will result the establishment of laws, by the aid of which it shall be possible, in other regions, to deduce facts of geological history from an examination of the relation of structure to drainage.

A LESS DIPLOMATIC CHALLENGE came two decades after Gilbert's paper, in an 1897 article in *Science* magazine by Samuel F. Emmons (1841–1911). He, like Gilbert, was one of the first geologists hired by the new U.S. Geological Survey. Emmons was also one of those fired in 1892 when Congress, in an attempt to force Powell to resign, slashed the budget of the Survey. They succeeded, and by the time Emmons wrote five years later, Powell's career and influence were over. Wallace

Stegner titles the chapter in which he describes this sad, final phase of Powell's life, "Consequent Drainage." The fall was a long time coming, but like that of the man with an Appointment in Samarra, fated. We owe Powell a few pages to understand why.

With publication of his *Explorations* in 1875 and, a year later, *Report on the Geology of the Eastern Portion of the Uinta Mountains*, Powell's direct contributions to his founding profession, geology, had come to a close. The next important volume in his writings reflected a change in his interest and ambition, one that would lead him to have more influence than any unelected person in the country (and more than most of the elected ones), but that would also bring about his downfall.

His years on hardscrabble Midwestern farms had made Wes Powell only too familiar with the farmer's lot. When he was only twelve and his brother Bram only ten, their father announced that they would have to work the land, as he would be off evangelizing full time. Even as a grown man in his thirties, Powell was slight, though in his later years he became portly. At age twelve, he was only a boy, yet the main responsibility for his family's welfare fell to young Wes, and he bore it. He had learned as only a farmer can the importance of having enough rain and at the right time. In his native Midwest, precipitation falls year-round, though in winter often as snow, and up to forty or fifty inches per year. One year might see a drought, but if so, over the next year or two, rainfall was likely to return to normal. But the pioneers on the Oregon Trail, the forty-niners, and, in Powell's day, travelers on the new transcontinental railroad, could not fail to notice that as one traveled west across the midsection of the country, rainfall declined, vegetation became sparse, trees nearly disappeared, and the sky became big. Meteorological data showed to the satisfaction of any objective person that the 100th Meridian provides a useful marker of adequate rainfall for farming to its east and inadequate to its west. This line of longitude bisects the Dakotas and Nebraska, passes through western Kansas to where Oklahoma shrinks to its panhandle, and crosses into Mexico at about Laredo. It separates a region to the east where rainfall exceeds twenty inches per year and one to the west where, except for the Pacific

Coast and a few isolated spots, it never does. Powell the farmboy and Powell the scientist knew that if it rained less than twenty inches, a farmer had to irrigate.

We could describe John Wesley Powell as an activist, though in the latter half of the century after him that term took on a pejorative cast. But if by it we simply mean a person unwilling to accept the world as he found it, unsatisfied unless engaged in trying to make it better, Powell would be activism's exemplar. Although, like any of us, he had complex motives for his actions, he was unwilling to leave his young country with a blank spot on its map the size of a large European nation. He was unwilling to leave geology in the hands of stale and stalled European theories. As he saw thousands of unprepared countrymen and new immigrants head west, encouraged by the free land provided by the Homestead Act of 1862 and by the lies of western boosters, he knew that for those who got west of the 100th Meridian, there would not be enough water. He was unwilling to abandon them to their fate. After all, the history of his own family had taught him that to leave hardscrabble farmers to fail, or worse, was inhumane. Powell first acted as a scholar: in 1878 he published a detailed 200-page treatise entitled *Report on the Lands of the Arid Region of the United States, with a More Detailed Account of the Lands of Utah*.

The years right after the Civil War, a decade before Powell's three major publications, turned out not to be a good time to argue the validity of the 100th Meridian as a meteorological boundary that ought to limit the ambitions of a young nation on the move. Starting in about 1865 and for several years thereafter, rainfall in the Great Plains registered well above its historic average, though in a region so recently explored as the West, the term "historic average" was dangerously misleading. Guides returning from wagon train trips to Oregon reported that western Nebraska, "usually blond from drought or black from prairie fires, had turned opalescent green." Some detected the hand of God intervening to provide for his chosen people. Those of a more scientific bent sought another explanation. To understand their theory, it helps to remember that throughout history people have been tempted to believe that human intervention can cause rain to fall. In the early nineteenth century, Napoleon, and later, Civil War veterans, believed that they had seen proof that cannon fire brought on rain. In the 1890s, Congress appropriated money to test whether explosions could bring moisture to parched West Texas; to conduct the experiments they hired the inaptly named R. G. Dryenforth. He flew sacks of dynamite aboard balloons and exploded them aloft. Sometimes it rained immediately; often it did not. Or, it poured but Mr. Dryenforth was nowhere around.

Thus, given the mind-set of the post-Civil War years, it was natural for people to ask what changes had been going on in the Plains that might have caused more rain to fall. To be sure, contending armies had not shot off barrages, but a different kind of battle had taken place: thousands of acres of soil that in all of earth's history had never felt a plow's blade now lay broken asunder. If one believes that cannon fire can bring rain, is it impossible that tilling the soil could do the same? Not when considerable scientific authority supported the belief. Professor Cyrus Thomas, a member of the Hayden Survey, intoned, "Since the territory [of Colorado] has begun to be settled, towns and cities built up, farms cultivated, mines opened, and roads made and traveled, there has been a gradual increase in moisture." So far, the good professor had merely made an empirical statement. But he went too far: "I therefore give it as my firm opinion that this increase is of a permanent nature, and not periodical, and that it has commenced within eight years past, and that as population increases, moisture will increase." Hayden himself endorsed the theory, citing evidence that planting trees increased rainfall. These men were not crackpots or believers in thaumaturgy, though there were plenty around who were and did: perhaps plowing the ancient soil released trapped water into the air; perhaps the smoke from the trains precipitated moisture; perhaps Hayden was right and newly planted trees were the reason. But whatever the cause, the empirical evidence was undeniable. Americans had begun to move west, and just when and where they needed it, more rain fell than in anyone's (awfully brief) experience. With politicians, newspaper editors, preachers, railroad men, land speculators, and hucksters of all sorts egging them on, "the tired, the poor, the huddled masses yearning to breathe free" needed no further persuasion. But Powell and his men knew it was all bunkum. He was conscience-bound to say so and try to prevent a grand national folly and countless human tragedies.

The Homestead Act had decreed that a settler would get 160 acres, a "quarter-section" of the standard measure of area: a block one mile on a side: a "section." Authorities said that a farmer needed only one-fourth of a section to support his family and would even have land left over. From this interpretation of how to bring about the Jeffersonian ideal of a nation of small farmers, many of the troubles of the West, right up to the twenty-first century, would devolve.

Beyond the Mississippi lay thousands and thousands of fallow quarter-sections; they needed only to feel the plow, and, as the recent increase in rainfall showed, even more rain would soon follow. Powell pointed out that, to the contrary, rainfall depended on large-scale climatic patterns. Human activity could not affect it for long, if at all. These patterns caused the East to be wet, allowing 160 acres to suffice amply for the family farm. Indeed, where a farmer could bring in two crops annually, even 80 acres ought to be enough. But in the West—say in such altitudinous states as Utah, Montana, and Wyoming—without irrigation, a quarter-section would support only four or five cows, a recipe for the failure of the small farmer and rancher. In the arid West, without irrigation, sufficiency required four full sections—2,560 acres.

Well, then, just irrigate the western quarter-sections, Powell said. But there wasn't enough water for that: all the water in the West from the Canadian border to the Mexican would irrigate only from one to three percent of the land. That meant there would have to be a lot of dry-land farms and ranches, each needing four sections. The dry-land farmer, though, needed protection from utter drought; an amount of water sufficient to irrigate twenty acres ought to insure him. But water was usually a long way from where a farmer would need it, and then it had a tendency to dry up. Even where water was plentiful and nearby, riparian law ruled in the West: those who owned the land adjacent to the water had exclusive rights to it and did not have to share. Nature and law thus conspired to provide some farmers and ranchers with all the water they needed, while leaving others with little or none. Somehow Nature would have to be improved upon so that the haves shared with the have-nots. But how to do so? By building a reservoir to impound the water and release it when and where farmers needed it. Who could do that job? Only the federal government had the capital and the necessary duty for the common good. Fair enough, but as if to show just how far he was willing to take idealism at the expense of practical politics, Powell went on to question why each farm and ranch, no matter how small, should be enclosed within its own fencing. Why not have a common exterior fence, inside which many farms would cluster together? After all, "It takes a lot more tin to make five eight-ounce cans than one forty-ounce can."

While he was at it, why not go all the way and question something even larger-scale and more sacred: the boundaries of the states? Anyone could see that mostly they made no sense. Some state lines did follow natural boundaries like rivers, but often as not they cut across rivers, mountain ranges, and every topographic feature, "sneering at natural reality." In an ideal world, state lines would reflect boundaries laid down by nature rather than the accidents of history and the self-interest of bygone politicians. In the West, where the boundaries of new states waited to be defined, the job could still be done right.

One can imagine the dull thud with which Powell's ideas landed. Supplying each farm with water for twenty acres, clustering farms and ranches together, drawing state boundaries according to topography rather than politics—like his later philosophical writings, even Powell's supporters could safely ignore such outré notions as merely the academic musings of an idealist and one-time professor of natural history.

Powell was an activist and an idealist at a time when the country was not in an idealistic mood, if ever it had been. The nation and its citizens wanted to get rich, their elected officials ought to help them do so—within the law, of course, even if that meant laws had to be recast and stretched. Having uprooted their lives and moved west, as everyone had encouraged them to do, who wanted to hear that rain was about to stop following the plow and that shortly the Great American Desert would reappear? Why should riparian water holders, who had ancient law on their side, share with anyone? The average farmer wanted to fence out his neighbor and his neighbor's livestock, not include them with his own in a giant commons. Politicians did not want to hear the dreaded word "arid" applied to their potentially fecund state, and it was they from whom Powell had the most to fear.

Congressmen did embrace the most practical and immediately beneficial of his proposals: the construction of a system of reservoirs across the West. Powell planned and set about this task as a scientist would: first, prepare a complete set of topographic maps for each state at appropriate scales between one and four inches to the mile. Second, study these maps carefully to determine the best sites for impounding rivers and sending the water from the resulting reservoir downstream to the irrigators. Thus, topography, which ultimately reflects underlying geology, rather than geography and politics, would show where to put the dams. The Survey began to map, but anyone could see that it would be a long, drawn-out project. Indeed, the USGS has never stopped mapping and remapping. Its "topo" maps became not only Powell's principal legacy, but the Survey's most important and best-known product. Geologist William Morris Davis went so far as to say that because of the Survey's maps, "The change from geographic barbarism of that earlier day to the relative civilization of today is due more to Powell than to any other one man."

But the westering nation of the 1880s and its elected officials did not have time for Powell's glacial timetable. A congressional resolution passed in 1888 that withdrew all arid lands from settlement until Powell's mapping was complete, which had the effect of suspending all land laws on such acreage, incensed nearly everyone. In a memorable passage, Stegner captures the mood of optimism, boosterism, ignorance, and animus that met Powell when he appeared before the House Appropriations Committee in June of 1890:

"Of course I have got a great respect for scientifically educated gentlemen, and I am always very much interested in their researchers and all that, but you can not satisfy an ordinary man by any theoretical scheme or by any science . . . One man can see in the ground no farther than another, unless there is a hole in it," said Moody of South Dakota. [Committee members] wanted to know who had defined the "arid region" and implied that it was a fiction of Powell's own, designed to get him extra powers. They doubted the necessity of his maps. Why couldn't the obvious reservoir sites be selected at once, a decent allowance being made for error, and the topographical survey be completed at leisure by the Geological Survey's crews? Also, what did he have to say to the fact that both Dutton and Nettleton, two of his experts, had testified that a topographical map was not necessary for selecting reservoir sites? What, they asked, did he know about the West? What did he know about South Dakota? Had he ever been there? When? Where? For how long? Did he know the average rainfall of the Black Hills?

Senator Moody [said] "Our people in the West are practical people, and we can not wait until this geological picture and topographical picture is perfected." Senator Stewart [of Nevada] had already made it personal. "Every representative of the arid region—I think there is no exception—would prefer that there would be no appropriation to having it continue under Major Powell."

The Congressmen also called Dutton to testify. He told them that although topographical maps were useful, they were "not indispensable." But as if to add insult to injury, Dutton went on to say, "The Director was under a misapprehension as to the degree of accuracy in his maps." Powell never forgave him. It is hard to imagine a more fundamental disagreement: Powell made the mapping his most important cause; Dutton questioned its necessity, legality, and accuracy. Stegner summed up the cost: "The tight loyalty of the bureau had been cracked, the Table Round had produced its Gawain, the Twelve their Judas." This seems far too harsh. Dutton had always doubted Powell's authority to use Irrigation Survey funds for topographic mapping, and made no secret of it. And he also testified that "no money has been better spent," suggesting that he was not disloyal but rather had his own opinion and stated it straightforwardly. Perhaps Dutton was no Judas, but that rare person, an honest man who honestly disagrees. But evidently he was as poor a politician as he was a superb writer. And unlike the Congressmen, Dutton was under oath.

By this time, influential members of Congress had grown fed up with the whiskery, unelected Major. Powell the idealist had gotten too big for his britches and now stood in the way of western expansion. They even accused him of wanting to dam the Grand Canyon, a bad idea that had not as yet occurred to anyone other than his detractors but that would before long. Powell had made insurmountable enemies in Washington. But worse, he had also made them outside the Capital.

EVER SINCE STENO, AS NOTED earlier, geology and paleontology have advanced together, progress in one leading to progress in the other. But by the nineteenth century, there was no field of science in which passions ran as high as in the study of fossils. As usual, among all remains, those of the fantastic dinosaurs evoked the strongest emotions. Any geologist who aligned himself (and almost all were men in those days) with one side of an argument about the terrible lizards ran the risk of being found guilty by association with those on the other side.

And arguments there were, culminating in the "Great Bone Wars" of the 1870s and 1880s, in which vertebrate paleontologists Othniel Marsh of Yale and Edward Drinker Cope of Philadelphia battled. The pair had begun as friends, collecting fossils together, but fell out, the story goes, when Marsh got Cope's excavators to send the fossils they unearthed to Marsh instead of to Cope. The two parted company forever after Cope published a description of a giant plesiosaur, only to have Marsh point out with undisguised pleasure that Cope had put the skull on the wrong end. Then began a great race to outdo each other in finding new specimens. As in a real war, their crews spied on each other, stole ammunition (the fossils), and blew up ammunition dumps (the excavation sites), rather than have them fall into enemy hands.

No geologist of influence could remain neutral. Cope had published with the Hayden Survey and always aligned himself with Powell's great adversary. Marsh had assumed the presidency of the new National Academy of Sciences on the death of Powell's mentor, Joseph Henry. Someone suggested that the Academy help sort out the four western surveys and eliminate redundancy. Marsh appointed an advisory committee that wound up not only espousing the consolidation of the four into one, but moving the USGS from Treasury to Interior and assigning land parceling to the Coast and Geodetic Survey. The NAS report "was identical with the program that Powell, Gilbert, and Dutton had been actively advocating . . . and almost wholly derivative from Powell's Report on the Lands of the Arid Region." King, with Powell's support, became the first Director of the recommended USGS, but after he resigned only two years later, Powell succeeded him. In 1882, Powell persuaded Marsh, still the president of the Academy, to lead the Paleontology Division of the USGS. This left no doubt which side Powell had taken in the Bone Wars. But, as in any battle, abandoning neutrality risks enemy fire. On January 12, 1890, as congressional souring on Powell had come to a head, The New York *Herald* opened with banner headlines:

> Scientists Wage Bitter Warfare. Prof. Cope . . . Brings Serious Charges against Director Powell and Prof. Marsh . . . Corroboration in Plenty . . . Allegations of Ignorance, Plagiarism, and Incompetence . . . Academy Packed in the Interests of the Survey . . . Will Congress Investigate?

The long article reflected a brew of hatred steeped during twenty years of rivalry. The main target of the blast was Marsh, but Powell suffered wounds both directly and by ricochet. Perhaps, because the rivalry did not consume him as it did others, and perhaps because it was his nature, Powell responded factually, without rancor, but with just the right touch of "kindly and ironic condescension." Marsh by contrast counterattacked with all guns blazing: "Little men with big heads, unscrupulous in warfare, are not confined to Africa and Stanley will recognize them here when he returns to America. Of such dwarfs we have unfortunately a few in science."

But then as now in Washington, even a charge fully answered and refuted leaves a public figure damaged. The charge appears on the front page and above the fold, the response somewhere in the back and below it. In Washington, D.C., black ink is often indelible. Powell had achieved great power and influence, but it rested largely on his personal integrity and the degree to which his agencies and programs delivered what members of Congress wanted. The Bone Wars left him damaged internally, though a few years would pass before the wounds became visible externally.

The president had appointed Powell, and only he could unappoint. But Congress held Powell's agency purse strings and could choke him off anytime they chose. And choke him off they did. The Senate voted to reopen the public domain, ending the possibility of careful planning for reservoir siting, and cut his appropriation for irrigation work. As Stegner put it, "The reduction of the Irrigation Survey from a comprehensive and articulated General Plan to an ineffectual and aimless mapping of reservoir sites was the major defeat of his life, and the beginning of the end of his public career." The defeat left Powell in charge of the Geological Survey, but that was still too much power for his enemies to tolerate. In 1891, with most of a vast country and its geological riches yet to explore, Congress delivered the first budget cut in the history of the Survey. Worse, the resulting appropriation set salaries and allocated the Survey budget to its different departments. Even the most obsequious administrator, and Powell was far from that, could not fail to see such a removal of the essential tools of leadership as a vote of no confidence.

Having wounded Powell and gotten away with it, a congressional *coup de grace* was inevitable. An amendment to reduce the Survey budget from $541,000 to $400,000 failed, but after "a little more horse trading," a narrow vote sliced it to $335,000. Powell's supporters got the sum back to $430,000, but that still required two paleontologists and fourteen others to lose their jobs. To add injury, the bill proposed to cut Powell's own salary from $6,000, the same that King had earned in

1879, to $5,000, but his friends defeated this final insult. Nevertheless, all too soon, "When the temple came down, the High Priest of Science was in it, a maimed man, in constant pain from the regenerated nerves of his stump, a man getting on toward sixty and in trouble with a wife who over the years had grown into something of a shrew. He was tired and he was licked."

On May 8, 1894, John Wesley Powell, surely one of the most able and dedicated public servants Washington, D.C., has ever seen, resigned. What a rapid and thorough fall for the man who as late as 1890 had "more sweeping powers in certain matters, than any man in the nation, not excepting the President." Soon the Senate confirmed his handpicked successor, Charles Doolittle Walcott, who went on to one of the most distinguished scientific and administrative careers of any American geologist. As we shall see in the next chapter, Walcott had proved his geologic mettle on the Kaibab Plateau. As for Powell, he stayed on at the Bureau of Ethnology until his death, where he tried nobly to help protect Native Americans from the onslaught of white settlement. The photographs of Powell exchanging information with Tau-gu and Tau-Ruv may well be worth many words in capturing Powell's attitude toward Indians. Influenced by his Methodist, abolitionist parents, and by his time at Oberlin and other liberal institutions, Powell saw Negroes and Indians, if not as equals, at least as fellow human beings. Contrast his attitude with that of another of the western Survey leaders, George M. Wheeler, who wrote, "While the fate of the Indian is sealed, the interval during which their extermination as a race is consummated will doubtless be marked . . . with still more murderous ambuscades and massacres."

Congress finally did launch Powell's Great Plan, in the very year he died, by passage of the Newlands Act that led to the Bureau of Reclamation. His humane idea wound up creating a bureaucracy that has left the West with almost no free-flowing rivers of any size, with vast subsidies to agribusiness to grow surplus crops at the expense of third-world countries, and with at least one national wildlife "refuge" turned into a selenium-poisoned abattoir. It is hard for this author to believe that John Wesley Powell, the lover of wild rivers since boyhood, the canyon runner, the frugal and responsible public servant, the champion of the yeoman farmer and the little man, would have approved.

Considering the fate of the men who headed the four great western surveys is worth another paragraph. As we have seen, Powell's congressional enemies broke him and retired him to a sinecure at the Bureau of Ethnology. Pained by this treatment, by his amputation, and by his shrewish wife, he still remained an optimist and a believer in the potential of America. Hayden, a man so self-absorbed he could never find time to write the obituary of his oldest and best (and almost his only) friend, died at age 59 of syphilis contracted from a prostitute. In 1879 the U.S. Army granted George M. Wheeler a disability discharge. "Embittered over rivalries with 'scientific filibusters' like Powell, his beloved survey abolished in favor of the civilian U.S. Geological Survey, Wheeler retired into anonymity, laboring for another decade over his accounts. By the time his Geographical Report appeared in 1889, few people cared about the survey or even remembered it." But what of Clarence King, he who showed more youthful promise than the other three put together? This friend of Henry Adams was not just at home in, but was the star of, the finest salons in America—a man for whom the phrase "brightest and best" might have been invented. King had begun to think of leaving the USGS almost before he arrived as its first director. Following a pattern that continues to this day, he quickly abandoned public service for the more lucrative private sector. The sybaritic King succeeded in amassing a fortune of nearly a million dollars in Mexican mining, but saw it "dissipated in years of indulgence abroad and annihilated in the Panic of 1893," requiring him to mortgage his art collection to his friend John Hay, Lincoln's Secretary and biographer. The low point came when police arrested King for disturbing the peace and he wound up in the Bloomingdale Asylum. When he died in 1901, King left behind a secret Negro wife and five children, who only then learned that the name of their husband and father was not "James Todd."

Powell was far ahead of his time in every area in which he had an interest. He had his faults, but what set him apart from his three rivals can be stated in one word: character.

Stegner provides the ideal eulogy: "Let us imagine Major Powell buried [beside the Missouri], perhaps seated on his horse like the Omaha Chief Blackbird. From the river bluffs, looking over the West that was his province, he can perhaps contemplate the truly vortical, corkscrew path of human motion and with some confidence wait for the future to catch up with him."

Chapter 10

Antecedence
in Doubt

〜〜〜〜〜〜〜〜〜〜〜〜〜〜〜〜〜〜〜〜〜〜〜〜〜〜〜〜〜〜〜〜〜〜〜〜

It is no surprise that not only did Powell fall out of favor with Congress, not only did he incur the enmity of Cope, Hayden, and their supporters in the Bone Wars, but his claim that the entire Colorado River system is antecedent also came under attack. Hired by Clarence King for his Survey, Samuel F. Emmons had never been a Powell man. To find oneself fired from the USGS, as Emmons did, while Gilbert, Willis, Walcott, and others were not, even though the reason was budgetary, must have rankled then as it does today. In less visible publications than *Science*, Emmons had taken exception with Powell by arguing that the Green River is not antecedent, but superposed. He began his 1897 article by noting that in his *Explorations*, Powell had said he would reserve the subject of the origin of the Green River for a fuller discussion in his forthcoming report on the geology of the Uinta Mountains. Yet when that report came out, the promised discussion was missing. Emmons snidely speculated that this was because "on further study [Powell] had found the difficulties in the way of this theory too great to be explained away." In fact, Powell's report did contain two paragraphs of explanation for his conclusion that the Green is antecedent, though they are not his clearest or most convincing. When parsed, his words remain little more than an assertion. It is easy to see why they did not satisfy the bitter Emmons. On the other hand, the terse Major evidently thought them sufficient:

Recurring again to the valleys of the Uinta Mountains, it may be well to remark here that, coming from the Rocky Mountains to the study of the Uinta Mountains, I at first supposed that the valleys of this region also were superimposed upon the rocks now seen, but gradually, on a more thorough study, the hypothesis was found to be not only inadequate to the explanation of the facts, but to be entirely inconsistent with them; and again and again I visited the region, and I re-examined the facts, and at last reached the conclusion which I have heretofore stated.

A brief reference to the character of this evidence may not be out of place here, though I reserve the subject for a more full discussion in my report on the geology of the Uinta Mountains. If the valleys were superimposed on the present rocks, they must be consequent to rocks which have been carried away; but the valleys consequent upon the corrugation, which was one of the conditions of the origin of the Uinta Mountains, could not have taken the direction observed in this system; they would have all been cataclinal [running in the direction of the dip of the strata], as they ran down from the mountains, and turned into synclinal valleys at the foot, forming a very different system from that which now obtains. Again, the later sedimentary beds, both to the north and south, were found not to have been continuous over the mountain system, but to have been deposited in waters whose shores were limited by the lower reaches of the range; that is they all gave evidence of littoral origin, and, further, that the principal canyons through the mountains had been carved nearly to their present depth before the last of these sediments were deposited.

Emmons summarized the facts on which he and Powell did agree. Uplift of the Uintas began in the Cretaceous Period, and, during the subsequent Tertiary (see the geologic timescale in Figure 5, page 112), 8,000 feet of sediment accumulated in lakes that "washed either flank of the range." But as emphasized in the last sentence of the previous quotation, Powell claimed that deep canyons existed even as sediments accumulated on the lake bottom. If the river is older than the lake, Emmons asked, "What, then, became of the river while 8,000 feet of sediment were being deposited? It could hardly have continued its course at the bottom of the Tertiary lakes while the sediments were depositing. But if it ceased to flow . . . its bed must have been filled with sediments . . . and when the lakes were finally drained . . . it is hardly conceivable that it should have attacked at exactly the same point it had entered before." Touché.

Emmons also claimed to have caught Dutton in an error. In a paper he wrote for one of the annual Survey reports, and which he repeated in his article in *Nature* magazine quoted on page 113, Dutton gives what Emmons calls a "most graphic" description of the course of the Green River through the Uintas. After describing in his inimitable style how the Green River cuts its way through the Canyon of Lodore, Dutton writes that then "a strange caprice seizes it. Not satisfied with the terrible gash it has inflicted upon this noble chain [the Uintas], it darts at it viciously once more, and entering it, cuts a great horse-shoe cañon more than 2,700 feet deep." Here Dutton's prose may have gotten the better of his science. For if he referred to the Horseshoe Canyon that Powell named, it lies to the north of the Uinta Range, not to the south where Dutton placed it.

As to what did decide the position of the modern drainage channels, Dutton said that they were "determined by the configuration of the surface existing at or very soon after the epoch of emergence. Then, surely, the water courses ran in conformity with the surface of the uppermost (Tertiary) stratum." This seems to say that the river appeared *after* the lake disappeared, contradicting Powell's previous conclusion. In all this, one senses that Powell and Dutton did not quite have their story straight, perhaps indicating that they were wrong about antecedence. Emmons did not go quite that far, but summed up with Gilbertian tact: "It would seem proper that the antecedent origin of this river should be held in abeyance until some positive evidence of it can be furnished." Touché again.

To get a century ahead of our chronological account, abeyance lasted until 1986, when USGS geologist Wallace Hansen, a Utah native and longtime student of the Green River and Dinosaur National Monument, published a report that appeared finally to settle the old question. Hansen's analysis is worth a moment because other geologists would espouse a similar process to explain the Grand Canyon.

The Green River heads due south through southern Wyoming and northeastern Utah, straight for the Uinta Range, but, as it nears, it veers off and travels some forty to fifty miles east through Brown's Park (see Figure 2, page 15). Then it turns abruptly ninety degrees southwest and plunges through the Gate of Lodore and into the deep canyon of the same name. Hansen discovered that Lodore Canyon did not appear until about five million years ago. Before that, an ancestral Green River had run southeast in a canyon through what is now Brown's Park, toward a young Yampa River. Erosion of the rising Uinta Range then buried the old canyon of the Green under hundreds of feet of sediment (the Brown's Park Formation). In some places, the original canyon floor lies at a depth of 1,200 feet below the present surface. Eventually the gravels so clogged the Green that they diverted the river to the south, across the Uinta Range, into which it has carved the Canyon of Lodore. Thus, Emmons was right: the present Green River is superposed, not antecedent.

But if the Green River, Powell's best example of antecedence, is superposed, the rest of the Green and Colorado river systems, including the section through the Grand Canyon, are also likely superposed. The possibility required analysis by a new set of geologists who bore no allegiance to Powell and Dutton. This requires us to shift our terrain from the Green River, where Powell first met white water and, as far as we know, first began to develop his ideas about river history and base level. From now on, we will move hundreds of miles downriver to the Grand Canyon.

1. John Strong Newberry in 1884, geologist with the Ives Expedition of 1857–58 and the first to explore the Colorado Plateau.

John Wesley Powell, age 31 (right), and brother William Bramwell Powell, circa 1865. J. W. Powell was soon to return to Illinois to become professor of Geology at Illinois Wesleyan in Bloomington.

The Second Powell Expedition Commences at Green River, Wyoming, May 1871

3. "Our First Camp." The men await Powell's arrival to begin their journey, May 4.

4. The expedition launches from the left bank about a half-mile from the Union Pacific Railroad bridge, May 22. *Left to right:* In the *Cañonita*, E. O. Beaman, Andrew Hattan, Walter Powell; in the *Emma Dean*, Steven Jones, John Hillers, John Wesley Powell, Frederick Dellenbaugh; in the *Nellie Powell*, Almon Thompson, John Steward, Frances Bishop, Frank Richardson.

5. Original wood engraving of the Gate of Lodore, from Powell's book *The Exploration of the Colorado River and its Canyons,* first published in 1875 by the Government Printing Office as *Exploration of the Colorado River of the West and Its Tributaries.*

6. Seventeen-year-old Frederick Dellenbaugh in the "Heart of Lodore," June 1871. The artist-photographer ran away from home to join the Second Powell Expedition. He became a well-known painter, world traveler, and founder of the Explorers Club.

7. Men of Powell's Second Expedition portaging through Hell's Half Mile, Lodore Canyon, May 1871.

8. Echo Park looking downstream, July 1871. Steamboat Rock, where Powell may have been saved by a pair of long johns, is on the right. From the left, the Yampa River flows smoothly in. The Bureau of Reclamation planned to dam this area in the 1950s. Conservationists successfully opposed the dam, but in a Faustian bargain, they agreed to the Glen Canyon Dam above the Grand Canyon. Echo Park is in Dinosaur National Monument.

9. Marble Canyon, August 1872, named by Powell for the prominent Redwall Limestone.

10. Kanab Creek, looking downstream about one mile above its confluence with the Colorado River in the Grand Canyon. Photo taken during the Birdseye Expedition of 1923.

11. Tau-gu, Chief of the Paiutes, with Powell, age 39, overlooking the Virgin River, circa 1873.

12. Powell in Indian dress and Tau-Ruv, a member of the Paiute Indian tribe. Photo taken in the Uinta Valley, eastern slope of the Wasatch Mountains, Utah, circa 1873.

13. The Powell-Ingalls Special Commission meeting with Southern Paiutes near St. George, Utah, September 1873. Powell stands at far left with empty sleeve.

14. Powell during a visit to Flagstaff, Arizona, circa 1891.

15. Clarence King, first director of the USGS, in 1879. He soon resigned to enter the mining industry.

16. The Hayden Party, first to ascend Holy Cross Mountain in Colorado, at mess in 1873. At far left, Ferdinand Hayden. At far right, painter William Henry Holmes.

17. Powell, circa 1896, two years after congressional enemies forced him from office.

18. Grove Karl Gilbert, circa 1898. A group of twentieth-century geologists would name him "America's Greatest."

19. Clarence Edward Dutton, the one Grand Canyon geologist who qualified as a poet, circa 1865.

20. The Grand Canyon, looking east from the foot of Toroweap, 1872. Likely this photograph was the inspiration for Holmes's *Panorama from Point Sublime.*

21. William Henry Holmes (1846–1933), painter of *Panorama from Point Sublime,* circa 1900.

22. *Panorama from Point Sublime* (1882) by William Henry Holmes, who accompanied Dutton to the Grand Canyon in 1880. The painting appeared in Dutton's *Tertiary History of the Grand Cañon District.*

23. Charles Doolittle Walcott, third Director of the USGS, during his directorship. He served from 1894 to 1907, when he became Secretary of the Smithsonian Institution.

24. William Morris Davis while on the faculty of Harvard University, where he served from 1879–1912.

25. Eliot Blackwelder while on the faculty of Stanford University, circa 1935.

26. Chester Longwell while on the faculty of Yale University, circa 1949.

27. Arthur N. Strahler while on the faculty of Columbia University, circa 1956.

28. Charles D. Hunt, master geologist of the Colorado Plateau, in mid-career with the USGS, circa 1945.

29. Edwin McKee on the trail from Supai, circa 1928.

30. Barbara McKee on Hermit Trail, Grand Canyon National Park, circa 1975. She stands on giant cross-beds of Coconino sandstone.

31. George Billingsley of the USGS, who has mapped more Grand Canyon rocks than anyone else, logging over 70 river trips and more than 3,600 miles on foot, high above Duebendorff Rapid, Mile 132.

32. Ivo Lucchitta, renowned geologist and author of the beautiful, eloquent book *Canyon Maker*.

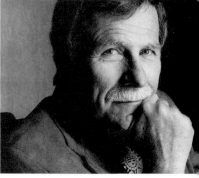

33. Richard Young, whose pioneering work on the Hualapai Plateau showed that the Grand Canyon had an equally deep ancestor.

34. Joel Pederson, one of a new breed of geologists trying to answer the old question: What caused the Grand Canyon?

FOR ALL THE REASONS THAT Gilbert explained so logically, the pioneer geologists found the Colorado Plateau a paradise, the ideal location from which to work out how erosion, transportation, and deposition shape the land surface. Collectively, they invented a new branch of geology called geomorphology, the study of the evolution of landscapes. But, as we have seen, Powell abandoned field geology for ethnology and administration; Dutton's curious mind led him into volcanology, seismology, isostasy, and all the way to the Sandwich Islands. When Gilbert could get out from behind a desk, his fieldwork took him to locations other than the Plateau—even to the moon. Gilbert's telescopic research prompted a congressional critic to say, "So useless has the Survey become that one of its most distinguished members has no better way to employ his time than to sit up all night gaping at the Moon." But Gilbert's ideas about the origin of lunar craters are close to today's accepted theory, showing just how far logic can reach.

William Morris Davis (1850–1934) picked up where Powell, Dutton, and Gilbert left off. He became not only the leading practitioner of the new field of geomorphology, but perhaps the best-known geologist in the country, partly because he went out of his way to write for and lecture to the general public. Davis began as a meteorologist but soon developed an interest in the origin of landforms. In 1876, he began teaching at Harvard. After only a few years, President Eliot warned him that his future at the university was in grave doubt. He offered Davis a reappointment, but accompanied it with the advice that in considering whether to accept, Davis should "look in the face the fact, that chances of advancement for you are by no means good." Whereas most would have withered in the face of a president who damned them with such faint praise, Davis took the promotion, earned tenure, and went on to become a legendary lecturer, professor, mentor, and author of over 500 publications.

By 1912, Davis had been professor of geology at Harvard for thirteen years. He was chair of the department and at the height of his power and influence, when suddenly, without explanation, he resigned.

The reason has always been a mystery. Fifty years afterwards, his son speculated that his father had not wanted to grow into an old and incompetent professor. But this explanation does not ring true, because for the two decades after leaving Harvard, Davis continued to do competently just what he had done as a professor: lecture to academic and popular audiences, conduct research, and write. Could it be that Davis wanted to show mighty Harvard that he and not the university had come to be the master of his fate?

In early 1934, a few days before his eighty-fourth birthday, Davis wrote to his younger son about his plans for yet another lecture tour that coming spring. His topic would not be his "crack piece" on the "Colorado Canyon," for that had been his subject on his most recent tour. Instead, he would speak about the geology and geomorphology of the eastern United States. Davis never got the chance, for a few days later he died.

An 1889 paper in *National Geographic* magazine entitled "Rivers and Valleys of Pennsylvania" had made Davis's reputation. In it he introduced a new paradigm: the geographical cycle—the passages in the life history of a landscape. The cycle begins with an original (consequent) river and ends when the river has "completed its task of carrying away all the mass of rocks that rise above its base level." Figure 6 illustrates Davis's geographical cycle.

In the first stage of the cycle, Youth, a fresh, new landscape first confronts the elements. Water begins to collect and to find its way into a meager set of small, consequent streams. Valleys have V-shaped profiles, falls, and rapids but have not had time to develop floodplains. Between the valleys lie broad, poorly drained divides with lakes and swamps. As erosion continues, the valleys widen and the ridges sharpen until almost every part of the land is in slope, consisting almost entirely of ridges and ravines. Systems of streams completely drain the land, so that "every drop of rain that falls finds a way prepared to lead it to a stream and then to the ocean, its goal." Now the landscape is Mature. Later, in Old Age, the topography comprises wide, open valleys, across which sluggish streams meander between barely existent divides. The

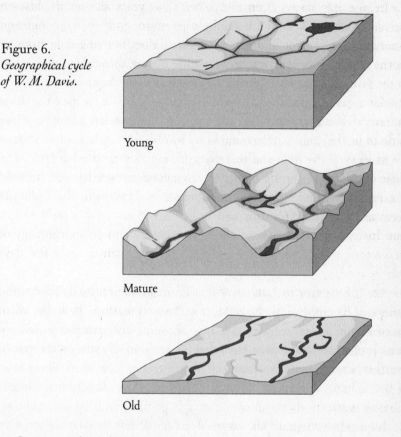

Figure 6.
*Geographical cycle
of W. M. Davis.*

Young

Mature

Old

surface is at a low elevation, near its base level, and, according to Davis, who was overly fond of anthropomorphic and Darwinian imagery, its mission is accomplished.

Like Powell, Davis recognized that a landscape could never actually reach base level, for as the elevation of the surface falls, the energy of the streams crossing it approaches zero asymptotically. In a letter, Davis used the term "peneplain"—almost a plain—to describe the more likely end product of his cycle. He and his followers found peneplains often in the geologic past, but, tellingly, nowhere in the present. The paradigm of a geographic cycle ending in a peneplain was to dominate the theory of physical geology for half a century.

In an earlier paper, Davis had drawn an analogy between a landscape and an oak tree. Few of us live long enough or, if we did, nowadays remain in one place long enough, to watch a single tree sprout from an acorn, become a sapling, rise to its mature height, grow old, and die. But by studying a succession of trees of various ages, from acorns to firewood, we can deduce the life cycle of a single tree. Davis used the same approach to understand the life cycle of a landscape. Any one far outlasts us, but our intellect allows us to observe landscapes at various points in their life span and thereby work out their entire life cycle. The putative ability of Davis's model to distinguish between young and old streams and their landscapes had obvious application to the Colorado River and the Grand Canyon.

OF THE EXAMPLES IN GEOLOGY that illustrate how reversible scientific fame can be, that of William Morris Davis is hard to beat. As often happens, but as few seem to take into account, his simplifying assumptions caught up with Davis. He based his geographic cycle on the postulate that when a fresh landscape begins to erode, it remains where it is, in an unchanging climate, its surface neither rising nor sinking, while it submits passively to its journey through youth, maturity, and old age. He acknowledged that this assumed an ideal, but he saw no need to inquire "whether or not any river ever passed through a single cycle of life without interruption." His model applied mainly to landscapes in humid climates and ignored the effects of glaciation, which have shaped modern landscapes in some parts of the northern hemisphere. We could not expect Davis to have known about plate tectonics, but Dutton's concept of isostasy and Gilbert's discovery of dynamic equilibrium ought to have figured more in his thinking. Stephen Pyne put it well: "For Gilbert, a graded stream was virtually timeless; for Davis, it marked the point in the evolution of a landscape by fluvial erosion at which a stream ceased to downcut and began lateral migration." In the end, Gilbert's concept of dynamic equilibrium supplanted Davis's geographic cycle.

To put it differently, not only do modern geologists not find peneplains, they reject the analogy of the oak tree. A tree remains in one place, in a constant climate, for all its life; landscapes do not. As Dutton showed had happened in the Colorado Plateau, landscapes certainly do not remain at one elevation for geological periods of time—they move up and down. Instead of reflecting a time sequence, landscapes display the effect of climate and erosion on different rock types during different degrees of uplift and subsidence. Changes take place when the climate changes, when the surface rises or falls, or when erosion exposes different rock types. Most fundamentally, unlike a tree or a human being, plate tectonics rejuvenates landscapes.

Between 1900 and 1902, Davis made three summer excursions to the Colorado Plateau and the Grand Canyon, spending a few weeks there each year. Even though his time on site was brief, the geologist's paradise ought to have been the ideal place for Davis to apply his concepts. Read today, his papers on the Grand Canyon show him to have been wrong in many respects, but also to have been one of geology's keenest observers.

Dutton had concluded that the Grand Canyon had formed in two cycles of uplift and erosion. The first occurred during the humid Pliocene Epoch and opened the canyon down to the level of the Esplanade, a broad bench that lies partway below the canyon rim, floored by the Esplanade Sandstone (see Figure 16, page 260). Davis called this the Plateau cycle. After a pause, uplift and consequent downward erosion began again, now under arid conditions, and started a second, or Canyon, cycle, during which erosion carved out the Inner Gorge, the deepest section, shown in Figure 16, eroded into Precambrian rocks. Davis spent most of one article presenting the evidence in support of Dutton's two-cycle model. At the end of the second cycle, he concluded, erosion had reduced the region to a vast peneplain.

As to the question on which the geologists of the day focused—antecedence or superposition—Davis said, showing as much tact as Gilbert and Emmons, that the pioneer geologists who had studied the Grand Canyon "might now modify in some degree the statements that they made thirty years ago . . . especially of the smaller streams." To Davis, the tributaries of the Colorado did not appear to be antecedent;

therefore, it did "not seem legitimate" to use them as evidence for the antecedence of the main stem. He called special attention, as has everyone who has studied the Colorado River, to the two great bends the river makes as it passes through the Grand Canyon: one around the Kaibab Plateau, and the other around the Shivwits Plateau (see Figure 1 and Figure 8, page 188). If the river were running before these giant structures existed, and thus had Powell's right of way, why did the river bend around them?

Davis's principal conclusion was that the Colorado River originated not on the floor of an Eocene lake, as Dutton claimed, but on a great mid-Tertiary peneplain, the evidence for whose existence Davis had discovered. The river had lowered itself onto the peneplain and then maintained its position through subsequent warping and faulting. Thus, it was not quite antecedent nor quite superposed, but a combination of the two, a possibility that Davis had foreseen in his *Rivers and Valleys of Pennsylvania*. Nonetheless, Davis did agree with Powell and Dutton that the Colorado began flowing through the region of the Grand Canyon in the mid-Tertiary. But, like Emmons, Davis rejected the strict antecedence of Powell and Dutton.

Through his lectures and writings, Davis went out of his way to make geology accessible to a wide audience. The message of his "crack piece" lecture on the Grand Canyon was surely the same as that conveyed in his articles for non-specialists. In one such paper, he sets out the true lesson of his beloved Colorado Canyon. The moral derives not from the immense width and depth of the Canyon, but from the thousands of feet of stratified rock exposed in its walls. The visitor must "soon perceive that the earth is necessarily not merely old enough to have allowed the erosion of the canyon in the plateaus, but old enough to have allowed, before the erosion of the canyon was begun, the formation of the huge pile of strata of which the plateaus are composed." In other words, Davis asks us to focus not just on the river's ability to cut through the great thickness of Paleozoic formations, but on the vastly greater amount of time it must have taken each of those strata to accumulate. This view takes us back to Hutton's unending rock cycle, to Dutton's marriage of subsidence and deposition on the one hand and uplift and erosion on the other, and to the abyss of time that they reveal.

Davis found the greatest lesson of all to derive from the Great Unconformity: the surface of erosion that lies between the Precambrian and the Paleozoic rocks of the Grand Canyon (see Figure 16), named by Powell. "Compare this vast buried plain, extending far and wide beneath the plateau strata, with the narrow canyon of the Colorado, and ask which one of these two pieces of erosional work was the greater; which one demanded the greater period of time for its accomplishment." Indeed, today we know that the Great Unconformity represents over one billion years of missing rock record, twice the length of all the time that has passed since the beginning of the Paleozoic, and vastly longer than the time it has taken to cut the Grand Canyon, even if that were to amount to fifty million years. Thus, again echoing Hutton, Davis found "one eternity ... recognized before another." He closes with a geologist's proper admonishment: "Some visitors foolishly spend part of their time, throwing stones from the plateau rim, to see where they will strike the slopes below the cliffs. A better use may be made of the rare privilege of standing in the presence of the awe-inspiring canyon by throwing one's thoughts into its marvelous depths of space and time, to see how far they may penetrate its wonderful meaning."

In the end, even had Davis's model not been flawed, it would have failed to measure up to the complexity of the Grand Canyon. Antecedence, superposition, and the geographic cycle were just too simple to explain North America's most complicated river and its canyons. Other processes, including one that Gilbert had mentioned in passing but to which no one had paid much attention, would prove more important. Powell's two models for how rivers evolve helped lay the path to understanding, but by the turn of the century their contribution had largely ended.

Anyone who delves into the history of science quickly learns that new theories, which better explain the evidence, or that avoid some fatal flaw of their predecessors, come to replace older theories. Some scientists have made careers out of provocative suggestions, most of which turn out to be wrong. Yet a well-thought-out proposal that raises important questions and suggests the line of research that might be most fruitful, even though it later proves wrong, is indispensable to the progress of science. In this spirit, Powell, Dutton, and Davis laid a vital foundation for Grand Canyon geology.

Chapter 11

The Same River Twice?

~~~~~~

Powell and Dutton differed with Davis, and, as we will see, with Powell's successor as Director of the Survey, Charles Walcott, on whether the Colorado River began on the floor of a great Eocene Lake or atop a peneplain and on whether it is antecedent or superposed. But they did agree on one thing: the river is old, at least as old as early Tertiary. Since that time, the Colorado has always followed the identical course. Until well into the twentieth century, everyone who studied the river and the Grand Canyon agreed.

The modern era of Grand Canyon geology could be said to begin with Eliot Blackwelder (1880–1969). In a 1934 paper entitled "Origin of the Colorado River," he was the first to suggest that the entire river is young. Powell and Dutton would have been taken aback to learn that, sixty-five years after the Major's maiden voyage, the river's age and history were still open questions. They would have been astounded to find that the origin of the Grand Canyon was the subject of a conference held in 1964, which reached consensus but not unanimity, and that yet another convened in the year 2000, with the same result.

But why should what had seemed to Powell and Dutton, if not quite to Gilbert, Walcott, and Davis, to have been settled in the nineteenth century still be under debate at the beginning of the twenty-first? One reason is that although the rock beds of the Colorado Plateau lie in simple, layer-cake stratigraphy, the Colorado River has proven to have an amazingly tortuous history. In other words, the history of the rocks and the history of the river have little to do with each other. Another reason is that even in paradise, deciphering geologic history is not easy. If erosion has not removed the rocks that interest you (and in the Grand Canyon district, it has done away with a pile of Mesozoic beds over a mile thick), other rocks often cover them up. The sedimentary rocks of the Plateau are colorful and eye-catching, but they usually contain neither fossils nor radioactive minerals, making their ages almost impossible to pin down directly. Folding, faulting, metamorphism, and other geologic skullduggery complicate interpretations, though less so in the Plateau than in mountainous regions like the Appalachians. A third reason is that by eroding away the sediments it previously deposited, a river devours its own progeny. This makes discovering its history like trying to trace the genealogy of an amnesiac orphan, all of whose identification and records are missing. The Colorado Plateau scientist is left to face the proclivity of Nature to leave only three kinds of evidence: circumstantial, negative, and none.

Yet perhaps another, more philosophical, reason explains at least part of the difficulty in determining the age and history of a river. Heraclites famously said, "You cannot step twice into the same river." We know that he was literally correct: rivers change constantly. If we return to a familiar river after a flood, almost everything, even the exact location of the river, may be different. But that is on our human timescale. Imagine the result when we add the changes to a river system over geologic time, as whole regions rise and sink, tectonic plates disappear down subduction zones, and continents collide. No one has found a Fountain of Youth for an oak tree, much less a human being, but tectonic uplift continually rejuvenates rivers.

As a river adjusts in the short run so as to balance each factor that affects it, as uplift gives it renewed downcutting energy over the long run, the very meaning of "the age of the river" begins to escape. Does "age" refer to the time since the river was a youthful, consequent stream on some long-vanished, primordial landscape; to the time since its last rejuvenation; or to the time of some other major event in its history? The question is as inherently hard to answer as it is to define the age of a rock that has passed through Hutton's cycle of sedimentation, consolidation, metamorphism, uplift, erosion, and sedimentation again, with "no vestige of a beginning, no prospect of an end." What is the age of such a rock? How old is a river?

For Powell, rafting the Green River to its junction with the Grand and on to the mouth of the Virgin, the last question would have seemed academic. Nothing could have been more natural for him than to think of the Colorado River as a single entity with a single life span. After all, he got on at one end and got off at the other, always on the same river. Otherwise, antecedence and its extreme, the Law of Persistence of Rivers, made no sense.

Geologists of the century after Powell's would realize that the history of the Colorado River is not so simple. Its path is anything but: in its long journey, the Colorado flows over at least three completely different types of terrain, crosses mountain ranges of different ages and types, wanders through broad valleys and incises deep into canyons, makes unpredictable bends, and now and again nearly loops-the-loop. Then, after the river has exited the Grand Canyon at Grand Wash Cliffs, the high jinks end, and the Colorado turns south and heads for the Gulf of California, as though at last it knows where it is going. Looking at the entire course of the Colorado in this way brings to mind not a mighty oak, but a plant assembled by grafting together unrelated limbs and branches. One of a macabre turn of mind, willing to risk offending the many lovers of the Colorado and the Grand Canyon, could go so far as to arrive not at an analogy with a handsome old oak, but to that ghastly creation of Baron Frankenstein, whose unrelated limbs and appendages were assembled in ways we do not wish to think about. Could the Colorado have been—one hesitates to use the word—stitched together from segments of other streams, each of them of a different age? If so, the question "How old is the Colorado River?" becomes even harder to pin down and antecedence and superposition even less up to the task of explaining the origin and history of the river.

But unless there is a process that can sew disparate stream segments together, the notion that a river could comprise pieces of other streams of different ages and histories is also merely an academic exercise. It comes as no surprise that Gilbert, in his primer on stream erosion in *Geology of the Henry Mountains*, was the first to identify the sewing machine. In a section called Unequal and Equal Divides, he pointed out that because the streams on the steeper side of a divide have greater velocity and energy, that side will erode more rapidly than the other. The inequality in grade causes the surface area of the steeper side to increase at the expense of the gentler. Thus, the location of the divide between the two sides migrates toward the gentler. As this process continues, at some point the divide has moved far enough for the drainages of the steeper side to intersect those of the gentler. Then the more energetic streams capture and reverse the direction of the waters of the less energetic. Where such stream capture happened in the recent geologic past, its effects may still be evident. But if capture happened scores of millions of years ago, and especially if a "great denudation" followed, it will be much harder, sometimes impossible, to recognize.

Though it may be difficult to define even what we mean by its age, in eroding its channel and depositing the resulting sediment downstream, a river does leave some evidence. One that has incised its way down through a succession of strata, as in the Colorado Plateau, obviously is younger than the youngest layer it cuts. Thus, in the Grand Canyon district, the Colorado must be younger than the Kaibab Formation, the 250 million-year-old limestone that forms the rim of the Canyon (shown at the top of Figure 16, page 260). In other words, the age of the Kaibab establishes the maximum age of the Colorado River. But because we know that erosion has removed thousands of feet of younger rock that once lay above the Kaibab Formation, saying that the river is younger doesn't get us anywhere.

In the 1930s, a few began to realize that they could turn the problem inside out: instead of showing where the river had flowed at a given time, they could show where it could not have flowed. For example, finding sediments incompatible with the presence of a large river would show that the Colorado could not have run through the area in question *at the time those sediments accumulated* (the italicized phrase will prove important later on). The Basin and Range Province to the west of the Grand Canyon, where steep but localized mountain ranges separate down-faulted basins, is full of such sediments. Figure 1 shows the location of the section of the Basin and Range Province that interests us. It lies west of a line running from the mouth of the Grand Canyon, near Lake Mead, northward to the Great Salt Lake. The province actually covers much of Nevada, western Utah, southwest Arizona, and southeast California, most of it not shown in Figure 1. Dutton said the Basin and Range Province resembled "an army of caterpillars crawling northward out of Mexico," and although he never saw an aerial view, Figure 1 bears him out.

Intermittent streams crisscross the isolated basins but do not leave them. The debris that washes down the slopes of the ranges thus piles up in the basins and remains there. Geologists can easily identify these rocks, for unlike the sediments deposited by a large river, they comprise only types found in the local mountains. Often an ephemeral lake forms in one of the basins; over time, it dries up, leaving behind telltale evidence of its short life in the salt and gypsum that precipitated out of its shrinking waters. Where we find interior basin sediments and evaporites, we can say with confidence that a through-flowing, major river did not exist at the time those deposits accumulated. A large river might have flowed before the interior basin deposits and then temporarily dried up, allowing those deposits to accumulate. After that, the climate became wetter and the river resumed its flow. But this kind of chronology is more complicated than the simpler history in which the through-going river arrived *after* the interior basin deposits had accumulated.

A river's own deposits precisely date the time of its existence—provided that we can find those deposits, recognize them for what they are, and measure their age. Geologists typically date rocks from their fossils or radioactive minerals, but as noted, many sedimentary rocks contain neither. To further complicate matters, large rivers, unlike interior basin deposits, amalgamate rock debris from scores and hundreds of miles upstream. These are the pieces of the rocks that the river happened to encounter in its downstream travels; they have a variety of ages that have nothing to do with the time the river flowed and deposited those pieces. A fossil of an organism that lived in the river, or near enough to fall into it and mingle with the river's sediments, would tell us when the river ran. Yet freshwater river gravels contain few hard-shelled organisms to leave fossils behind. Those remains that are present, like the sediments in which we find them, could have washed down from upstream deposits and be older than the river.

But before geologists can do anything, they have to find and recognize deposits left by the river. The more vigorous the river, the less evidence it leaves. The fast-moving Colorado picks up and transports downstream any incipient deposits; it once carried them to the Gulf of California, but today uses them to fill Lakes Powell and Mead. The sandbars on which Major Powell and his men camped on their first voyage not only likely contained few if any hard-shelled organisms that could provide fossils for geologists of the future—by the second trip, the bars themselves may have been replaced by new sediment. As we will see, several theories for the origin of the Grand Canyon hang in a kind of limbo because geologists have never found the critical sedimentary deposits that would confirm the theories. This indeterminate result leaves geologists uncertain whether the sediments never existed in the first place (evidence of absence) or whether they have simply not looked hard enough (absence of evidence). And the Colorado Plateau is a mighty big place.

AT THE TIME ELIOT BLACKWELDER wrote about the Grand Canyon, he chaired the Geology Department at Stanford University, a post he held for twenty-three years. Blackwelder had entered the University of Chicago intending to major in classics, but a senior-year course with noted geologist R. D. Salisbury, and two field trips to the Rocky Mountains, turned him toward geology. In 1903, after Blackwelder had taught for two years at his alma mater, Bailey Willis invited him along on a geological trip to China, sponsored by the Carnegie Institution of Washington. They traveled east on the new Trans-Siberian railroad; when in China they covered some 3,000 miles in that nearly unvisited country by just about every known mode of transportation. What stories Willis must have told Blackwelder about the pioneer Western geologists, all of whom he knew personally.

John Wesley Powell and Charles Walcott were among the last scientists to achieve national prominence without a college degree. Two generations after them, Blackwelder was one of the last non-specialists. His publications covered fossils, climate, structure, sedimentary rocks, Precambrian rocks, geomorphology, and petroleum geology. Blackwelder wrote in a simple and direct style, raising fundamental questions that in hindsight seemed obvious, but that no one before him had recognized or put so clearly. His memorialist said that his papers "have become classics in geologic literature, less for their originality or profundity than for the crystal clarity with which they illuminate subjects that previously had seemed bogged down in controversy."

In his *Origin of the Colorado River*, Blackwelder observed, "In the last half century, few geologists seem to have studied or considered the river as a whole." He instead took a bird's-eye view of the entire Colorado, from its ultimate headwaters as the Green, high in the Wind Rivers, to its merger with the Gulf of California:

> Rising in the high mountains of Wyoming and Colorado, [the Colorado River] traverses a series of wide basins, each of which seems to be an entity almost unrelated to the others. It cuts through the Uinta Mountains and the Colorado Plateau in deep canyons and repeats the act on a smaller scale several times between the mouth of the Grand Canyon and the Gulf of California. It runs nearly south for hundreds of miles, then for no obvious reason turns abruptly west, crosses northern Arizona, and again turns due southward in an erratic course. It enters the long Salton trough, the southern part of which is occupied by the Gulf of California, not at the upper end of the trench but at one side; and it shows its lack of genetic relation thereto by building a delta out into the trough, thus forming the basin which is now occupied by the Salton Lake.

If Powell, Dutton, and especially Davis had been correct about the age of the Colorado, it should show the features of old age that Davis had defined. But Blackwelder recognized that this "anomalous stream" does not show such features:

> The profile of the Colorado is not that of an old sedate river of long and normal history. It has rapids in its lower course and suffers many changes in gradient. It has not yet developed a meandering channel and a wide flood plain, except at intervals where weak rocks favor unusually rapid erosion.

If the Colorado River is old, as Powell and Dutton thought, one might find ancient deposits that would mark its passage. Indeed, the pioneer geologists, starting with Newberry, believed that they had found those deposits in the sediment-filled structural basins that lie along the lower Colorado River, south of the Grand Canyon, between the Grand Wash Cliffs and the Gulf of California (see Figure 1). Newberry wrote, "Doubtless in earlier times [the river] filled these basins to the brim, thus irrigating and enriching all its course." But Blackwelder believed that the deposits of the lower Colorado River are no older than Pleistocene, the recent Ice Age epoch. The beds that the pioneers had supposed to be Tertiary river sediments in fact contain only local rock debris; by the previous reasoning, they must be interior basin deposits. The presence of these rocks, which include extensive salt beds, shows that no big river flowed through the region at the time they accumulated. "For hundreds of miles east and west [of the lower Colorado], the only deposits of late Tertiary age . . . are laid down in desert basins rather than upon the flood plains of long rivers." Thus, Blackwelder could find no evidence that in the Tertiary a great river had existed where now we find the last stretch of the Colorado. This led him to ask the previously unaskable: "Did the Colorado River exist anywhere in Pliocene time? If not, how and when may it have come upon the scene?" Blackwelder concluded that the Colorado River was much younger than anyone before had thought.

If the Colorado were an old river, it should have displayed a full set of tributaries. The streams themselves might have withered as the climate subsequently turned arid, but the valleys they once occupied ought still to be present. Yet instead of a network of now-dry tributaries, Blackwelder found only desert basins. He saw no evidence that prior to the Pleistocene, a major stream had flowed through the lower Grand Canyon district. (As we will see, Richard Young later found such evidence.)

Blackwelder next asked another obvious question: when and from where did the Colorado River obtain the *sine qua non* of any stream: its water? The Colorado flows for a thousand miles through a desert. In its midsection—from the merger of the Green and Grand to the Virgin—a stretch of several hundred miles, it has only a handful of major tributaries (see Figure 1 and Figure 8, page 188). Thus, the water of the Colorado River must come from far away, from some "special circumstance of topography." Blackwelder identified that circumstance as the presence of the Rocky Mountains, with heights above 10,000 feet, to the east of the Plateau. As they ascend to pass over the High Sierras, winds blowing eastward from the Pacific lose much of their moisture. Were it not for the Rocky Mountains, these dry winds would descend and continue east at the same altitude. But instead, the Rockies force the wind currents to rise once again. In the process they cool and lose the last of their moisture as the rain and snow that fall on the western slopes of the Rockies and feed the Colorado.

Without the Rocky Mountains, there would be no Colorado River and no Grand Canyon. Thus, prior to the uplift that created the Rockies, the river could not have existed. Blackwelder saw evidence that the final uplift of the Rockies, which brought them above the elevation necessary for a perennial snowcap, had happened relatively recently. He concluded that all the evidence pointed to the Colorado's being an exceptionally young stream—surely one of the youngest in North America. Though today we know that the mountain-building event that created the Rocky Mountains began further back than he thought, Blackwelder's work was an important milestone in the history of Grand Canyon geologizing. Even if wrong in some respects, he was the first to refute the conclusions of Powell, Dutton, and Davis and to argue that the entire river—not only below the Kaibab but above it—is young.

CHESTER RAY LONGWELL (1887–1975) was the first to try to set both a maximum and minimum age for the Colorado. He had delayed his college entrance by several years to teach school and work at various jobs, but when he began he went all the way to a Ph.D. at Yale and from there to special prominence in American geology. Longwell was born in Mark Twain country in northeast Missouri, "an association that colored his speech, his many anecdotes, and perhaps even his way of thinking about life." His thesis topic was the Muddy Mountains of southern Nevada, west of the present mouth of the Grand Canyon and north of present-day Lake Mead (see Figure 1 and Figure 11, page 199), a region then unmapped topographically, much less geologically. Longwell spent the summer and fall of 1919 there, traveling by mule or horse and camping with hermits, Indians, and prospectors. What appealed to him was "the lure of the unknown," an apt description of almost everything about the Muddy Mountains, particularly their geology.

Longwell went on to join the Yale faculty and become recognized not only for the excellence of his science, but for his work as a mentor and especially, carrying on a Yale tradition, as the coauthor of one of the most widely read introductory geology textbooks. Eight of his Ph.D. students followed him into the National Academy of Sciences. The Geological Society of America elected five of them, as they did Longwell, to its presidency. Five received the Penrose Medal of the GSA, American geology's highest honor.

Longwell later conducted the original geological survey on the site that would hold Hoover Dam, work that required him to consider the history of the Colorado River. He wrote one paper on the geology of the region in 1936 and another in 1946, entitled "How Old is the Colorado River?" Longwell noted that although Blackwelder had argued that the river is young, geologist Charles B. Hunt, of whom we will hear more later, had argued that it is old. Furthermore, to Longwell's chagrin, both had cited his own work in support of their opposite conclusions. By 1946, he thought it was time to clarify his position.

In Longwell's field area, just west of the mouth of the Grand Canyon, the Colorado River has incised canyons through hard Precambrian rocks, like those of the Inner Gorge of the Grand Canyon. These canyons separate down-faulted basins filled with thick sequences of sedimentary rocks. It was the narrowness of Black Canyon, where the vessel of pessimistic Ives struck the fateful rock, that led the Bureau of Reclamation to choose it as the site of Hoover Dam. Longwell confirmed Blackwelder's observation that nowhere in the area west of the mouth of the Grand Canyon are there old river gravels of the kind that a major river would leave. Instead, the rocks comprise the alluvial fan debris, gypsum, salt, and limestone typical of the interior basin deposits that we find in the Basin and Range Province today, where indeed there are no large, through-going rivers.

As far as Longwell could tell, the youngest rock unit in the area is the Muddy Creek Formation. It crops out just west of the Grand Wash Cliffs, (see Figure 8, page 188 and Figure 9, page 193). The Muddy Creek has turned out to be the most critical formation in determining the origin and history of the Colorado and the Grand Canyon. The evidence that it is an interior basin deposit is so strong that according to Longwell there is "no possibility that the river was in its present position west of the Plateau during Muddy Creek time." Thus, the Colorado River, at least west of the Grand Canyon, is younger than the Muddy Creek Formation. Assuming that is, that an an older river did not first dry up and then resume its flow *after* the Muddy Creek sediments accumulated. But no one knew the age of the Muddy Creek Formation. Longwell could say only that it was "doubtful Pliocene," a conjecture that rested on a flimsy correlation with another similar-looking deposit that happened to contain Pliocene mammal remains, illustrating the difficulty of dating river deposits using fossils. By the time of his 1946 paper, Longwell's doubt had increased; now he thought that the scanty fossils in the Muddy Creek suggested an older, Miocene date. The age of the Muddy Creek would likely set the maximum age of the Colorado River, but in the absence of good fossil evidence and before radiometric dating, the age of the formation was unknowable.

Since the epochs, periods, and eras of the geologic timescale (see Figure 5, page 112) were defined, the Miocene has always been the epoch of the Tertiary Period just below and therefore just older than the Pliocene Epoch. But as geologists have studied more rock sections and dated them using better methods, they have had to shift the estimated ages of the boundaries between the epochs and periods. In Longwell's day, geologists estimated that the Pliocene began 13 million years ago and lasted until 1.8 million years ago. In the 1970s, based on better age dating, they redefined the beginning of the Pliocene at 5.3 million years, seriously shrinking the length of the epoch and making it confusing to compare the conclusions of geologists made before the boundary was reset with those made after. Our concern is that Longwell could not pin down the age of the Muddy Creek but had begun to think it was older rather than younger, moving the maximum age of the Grand Canyon back in time.

What about the other way of gauging the age of a river: by its own deposits? Resting above the Muddy Creek Formation, and therefore, by Steno's Law, younger, Longwell found beds of rounded pebbles left by a large river and composed of rock types from upstream, arranged in a way that allowed him to determine the direction of the river's current flow. These deposits he attributed "unmistakably" to the Colorado. Thus, he had located the time at which the Colorado began to flow through the Grand Canyon, but how to date that time? Again, the scarcity of fossils stymied Longwell; all he could come up with was the femur of an extinct camel, vaguely identified as "not far from *Camelops huerfanensis*." Longwell "pronounced" this specimen Pleistocene, but vertebrate paleontologist George Gaylord Simpson said the bone could be either Pliocene or Pleistocene. Whatever its age, the specimen might have been older than the river and have washed into and mingled with river sediment, an inherent problem in dating stream deposits.

In spite of his careful work, Longwell was unable to pin down when the Colorado River began to flow through the Grand Canyon. But he did show what it would take: knowing the age of the Muddy Creek Formation, which would likely fix the maximum age, was critical. If his tentative age identifications were right, the Colorado had not flowed through the Grand Wash Cliffs and into its lower course until at least the late Miocene, or even the Pliocene. Either would make it far younger than Powell, Dutton, and Davis believed, though not as young as Blackwelder thought. Longwell's work showed how desperately geologists needed to find a way to date fragmental sedimentary rocks that have no fossils.

In his 1936 paper, Longwell summed up with a few prescient

sentences that, nearly seventy years later, still serve well. Each of the current theories for the origin of the Grand Canyon corresponds to one of the possibilities he describes. Note that Powell's antecedence does not make Longwell's list:

> Thus, there is much that is unknown concerning the early history of the Colorado. The original stream may have started cutting on a high-level surface that overtopped all of the present ranges lying athwart its course [*superposition*]; or, integrating basins that became connected along the course of the river may have prepared part of the valley in advance [*lake integration*]; or, a stream emptying into the Gulf of California may have extended its valley by headward erosion, starting the cutting of the canyons and finally integrating with drainage from the Rocky Mountains [*headward erosion*]. Whatever the exact method of its initiation, it is probable that the river is following its original consequent course, which it has maintained by cutting canyons through hard-rock ridges and plateaus that have grown in height as the weak basin sediments adjacent to them have been etched away [italicized words added].

A decade later, Longwell had begun to wonder whether Blackwelder, who thought the river is young, and Hunt, who thought it is old, might not both be right. Longwell noted that upstream from the Grand Canyon—for example, in the famous Goosenecks of the San Juan River, one of the few major tributaries of the Colorado River—meander belts incise into canyons. Meanders are evidence of a river that is at least mature; deep canyons are signs of youth—put them together and you have evidence of a complex history, far more so than Powell, Dutton, and Davis thought.

Thus, Longwell began to get the idea that different sections of the Colorado River might be of different ages. He speculated that above the Grand Canyon, east of the Kaibab Plateau, the river might have originally flowed to the south. Later, volcanism and uplift diverted it to its present westward course. This would make the downstream section, through the Grand Canyon, younger than the upstream section above the Kaibab. Longwell put his point somewhat mysteriously: "The suggestion of an earlier drainage southward from the Plateau does not rest on any direct field evidence known to me. However, in the field of speculation there are logical possibilities other than the persistence of a drainage pattern from Eocene time to the present. Evidence found west of the Plateau strongly suggests an alternative possibility." In the preceding long quotation, he lays out two such possibilities: lake integration and headward erosion.

Longwell's suggestion of a different course for the Colorado above the Grand Canyon was little more than a hunch, but the hunches of a geologist of his stature were well worth paying attention to. Sure enough, two decades after his 1946 paper, geologists returned to his idea that the ancestral Colorado might not have flowed to the west over the course of the present Grand Canyon, but instead ran south, toward the Rio Grande and the Gulf of Mexico. Later, a younger, western, downstream section worked its way headward and captured the older, south-flowing, upstream section. If this model is correct, the critical area lies at the putative point of capture: where the river makes a giant bend as it crosses the Kaibab, shown in Figure 8.

⬛

IF SEGMENTS OF THE COLORADO RIVER might have different ages, geologists also needed to look upstream in addition to considering the section west of the mouth of the Grand Canyon, as Blackwelder and Longwell did. This takes our story not only upriver, but back several decades in the history of geology.

More than two hundred miles east of the Grand Wash Cliffs and the Muddy Creek Formation, the Colorado River makes its way south and west through the southeastern corner of Utah, as shown in Figures 1 and 8. Beyond Lake Powell, Glen Canyon Dam, and the mouth of the Paria River at Lee's Ferry, the Colorado next passes the Paria Plateau on its right. It continues in a more or less straight line through Marble Canyon, heading directly for the Kaibab Plateau, the highest region adjacent to the Grand Canyon itself. Powell adopted the name Kaibab from the Indians, saying that it translated to "mountain lying down." The river makes a great bend around the Kaibab and then proceeds northwest to the entry of Kanab Creek, where it turns southwest again. It stays on this course for several score miles, heading straight for the next large plateau, the Shiv-wits. It sidles up to the Shivwits and edges alongside it for thirty miles, only to make a huge, obtuse bend around that plateau as well. Even if one knew nothing else about the geology of the Colorado River, the two great deflections make it hard to escape the conclusion that by deferring to the underlying geological structure, the river must be the younger.

The Kaibab Plateau is the first and largest obstacle in the path of the Colorado after the Uintas; explaining how the river got across it has always been the key test of any theory of the origin of the Grand Canyon. The first person to study the Kaibab in more than a reconnaissance fashion did so at the end of the nineteenth century, decades before Black-welder and Longwell. Charles D. Walcott (1850–1927) was at the time a brand new employee of the still new U.S. Geological Survey. The education of this 32-year-old junior paleontologist did not extend even to the perfunctory college attendance of John Wesley Powell. Yet Walcott went on to become not only an outstanding geologist and paleontologist, but Powell's successor as Director of the U.S. Geological Survey, Secretary of the Smithsonian Institution, and President of the National Academy of Sciences. He discovered the weird Cambrian fossils of the Burgess shale in Canada, about which Stephen Jay Gould wrote in his *Wonderful Life*.

Founding Director Clarence King had hired young Walcott in 1879 as USGS employee number 20. His salary was $50 per month, the same as Gilbert's wage as a "volunteer" for Newberry some 25 years before. Walcott's first assignment in the Grand Canyon was to "measure section": to determine the identity and thickness of each sedimentary rock forma-tion and thus to define the geologic column for the region. And what a section it was, stretching from the Eocene beds exposed near Bryce Canyon down Kanab Creek all the way through the Cambrian layers outcropping at river level. Walcott and his crew worked from mid-August 1879 until mid-November by using a hand level, chain, and altimeter. They covered 80 miles laterally and 13,000 feet of vertical section. Age measurements made in the next century would show that Walcott's section comprised 500 million years of earth history. It might be the longest continuous geologic section ever measured, there being few other places where two-and-a-half miles vertically are accessible in one 80-mile stretch.

During the winter of 1882–1883, Walcott moved his field camp upriver to the Kaibab area, where he inevitably had to confront the antecedence theory. One of its two authors, Powell, was now his boss, Clarence King having departed the Survey; the other, Dutton, was a distinguished senior member and would become author of the definitive work on the Grand Canyon. Walcott published the first of his two papers on the Grand Canyon, which took on antecedence directly, in 1890, and the second in 1895, by which time he had succeeded Powell as Director of the Survey.

On its east and west sides, a stairstep flexure, which Powell called a monocline, bounds the Kaibab, as shown in the Major's cross-section (Figure 3, page 48). Walcott thought that the reason the rocks of the Kaibab had bent but not broken was the weight of the "thousand feet or more" of Mesozoic rocks which had once existed above them and had pressed down upon them and held them in place. As to the question of how the river got across the Kaibab:

> It is difficult to understand how the canyon could have existed even to a limited depth, in its present position, at the time of the elevation of the Kaibab Plateau. An explanation more in accord with observations . . . is that while the uplifting of the plateau and the East Kaibab displacement [monocline] were progressing, the Colorado River was cutting its channel down through the Mesozoic groups that then rested on the Paleozoic rocks in which the present canyon is eroded, and that, instead of cutting a channel down through the limestones and sandstones of the Paleozoic, as the plateau was elevated, it was cutting down through a fold in the superadjacent Mesozoic rocks.

Walcott thus concludes that the Colorado River across the Kaibab section is superposed: let down through an overlying sedimentary cover. The river would then be not only younger than the Kaibab structure, but also younger than the Mesozoic rocks that once capped the arch. Showing considerable tact of his own, Walcott wrote that this new interpretation "would probably necessitate some change in the now accepted views concerning the manner of erosion of the broad outer and narrow inner canyons, west of the Kaibab division of the Grand Canyon."

Walcott's analysis showed that Powell's three-way classification of streams and his conclusion that the Colorado River is antecedent were too simple, if not simply wrong. As the river eroded its way down through a sedimentary cover, a process that Powell would have called superposition, the Kaibab structure rose, which would have made the river antecedent. Obviously, one river can combine the two processes, even at the same time, as Davis had recognized.

The Kaibab structure is an anticline, a rock upfold, an arch. Instead of being level, however, the crest of the arch slopes downward and in effect "runs into the ground," as does the ideal anticline shown in Figure 7. This causes the beds to outcrop in a set of nested, arcuate bands. Walcott believed that on the south side of the Kaibab Arch, the now-vanished Mesozoic rocks would have outcropped as shown. As the river lowered itself down through the succession of Mesozoic cover rocks, it became stuck in a curving band of easily eroded shale, locked in place by bands of harder rock on either side. Continuing to cut down, the river entrenched itself in this easily eroded belt. Today, the giant bend the river makes around the mountain-lying-down preserves the arcuate shape that it inherited from the southward-plunging Kaibab Arch.

After the papers of Walcott and Davis, nearly half a century would pass before anyone would again study the Kaibab in detail. In two papers in the mid-1940s, Arthur N. Strahler (1918–2002), a professor at Columbia University who later became a prolific writer of excellent textbooks, took the Kaibab story a good deal further. Strahler began by praising Walcott's outstanding geology, done a half-century before, which he said was so

Figure 7.
A plunging anticline, the geologic structure of the Kaibab Plateau.

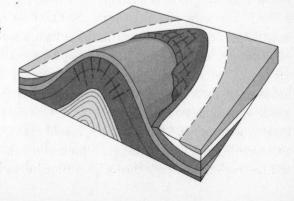

thorough it needed no repeating. Strahler was not merely being polite.

Strahler conceived of two different ways in which the river could have reached its present level, and illustrated one of them with three block diagrams. Two decades later, a group of experts at the first conference on Grand Canyon geology would adopt those diagrams to outline their own theory. Their version is shown as Figure 10 on page 195.

As one possibility to explain how the river came to cut a deep canyon across the Kaibab structure, Strahler envisioned a small stream flowing across the buried arch from east to west. Faulting then dropped the region to the west of the Colorado Plateau, creating the Basin and Range Province and lowering the base level of the small stream, which "stimulat[ed] it to conquests in its upper part." In other words, the drop in its base level energized the small stream so that it cut downward and upstream much more vigorously, the process of headward erosion that Longwell had mentioned. The second possibility was Walcott's model: that erosion had gradually let down the Colorado, already a major river, through the Mesozoic cover rocks and onto the Kaibab Arch until it struck and became fixed in a curving band of weak shale. In the first theory, the river approached the Kaibab by working its way backwards and uphill, from the side, like a burglar coming in through the window. In the second, the river descended from above, by cutting a hole in the roof. In either case, the river is younger, possibly far younger, than the Kaibab arch.

Though Walcott and Strahler did not settle the matter, both concluded that across the Kaibab, and possibly upstream from there, the Colorado River might be as old as early Tertiary—Eocene or even Oligocene. Blackwelder later disagreed, but both he and Longwell had demonstrated that where it leaves the Grand Canyon, the Colorado River is no older than latest Miocene. Geologists confronted at least a paradox if not a physical impossibility: a river whose upstream and downstream sections differ in age by tens of millions of years. And as Longwell had summed up, they had at least three possible explanations: superposition, lake integration, and headward erosion. The history of the Colorado River and the Grand Canyon were proving far more complicated than Powell and Dutton had anticipated. It was time for the generation who would succeed Blackwelder and Longwell to take their turn at answering the great question.

PART THREE

A New Theory

Chapter 12

Paradox

〜〜〜〜〜〜〜〜〜〜〜〜〜〜〜〜〜〜〜〜〜〜〜〜〜〜〜〜〜〜〜〜〜〜〜〜〜〜

By the 1950s, a century had passed since Newberry first visited the Colorado Plateau. Yet the answers to the most obvious questions: "What caused the Grand Canyon?" and "How old is the Colorado River?" seemed no closer. The brightest and best American geologists of the founding century of their science had failed to find the answers. The Colorado Plateau had seemed to Newberry, Powell, Dutton, and Gilbert the perfect place for geologizing. But plateau geology had instead turned out to be more like one of its maze-like areas: easy to get yourself into but difficult, if not impossible, to get out of. Indeed, a century of study had brought geologists to a paradox: the lower section of the Colorado River appeared to be millions of years younger than the upper.

The pioneer geologists of the Grand Canyon would have thought that having the downstream section be younger than the upstream was exactly the opposite of the way rivers behave. Dutton clearly set out their view:

> We are now in a position to trace the origin, growth, and history of the Colorado River, if not from the beginning, at least from an epoch near its beginning. Its creation was not the event of one epoch, but a gradual process extending through several periods. *The lower course, extending from the mouth of the Virgin to the Pacific, is the oldest portion* [italics added], and makes its appearance in geological history

a little before, but very near, the middle Eocene. There can be little doubt that in the middle Eocene the outlet [of the giant lake he had identified] was the lower course of the Colorado.

That the lower stretches of a river should be the oldest fits with common sense, for it is there that we find wide floodplains and meander belts, which take a long time to develop. And because as they age, streams erode headward and lengthen their channels, upstream sections of a river ought to be younger than downstream. How could the lower section be the youngest? Before geologists were ready to tackle that question, they had to hear from a formidable contemporary, a master of Colorado Plateau geology who had his own theory to propose.

CHARLES B. HUNT (1906–1997) wrote the first comprehensive review of the entire Colorado Plateau. Hunt had graduated with honors from Colgate University in 1928 and entered the doctoral program in geology at Yale. As an undergraduate and graduate student, he served three seasons as a USGS field assistant in Montana, Utah, and New Mexico and, in common with most geologists, found that fieldwork appealed to him. In those days, getting a job with the USGS required one to pass the most difficult geology examination in the country. Hunt did pass and elected to leave Yale early. In 1930 he began work for the Survey, mapping the coalfields of eastern Kentucky, a depressed region even in the absence of a national depression. Soon he transferred to the Colorado Plateau, his first love. In 1940, a world war already under way, the Survey transferred him to the search for manganese, a metal critical to making the high-quality steel that modern armor required. At the time a poor and unconfident writer, Hunt used the enforced inactivity of a bout with mumps to study up on technical writing. The effort must have worked, for he wrote many fine articles and several books. He often illustrated his writings with handsome pen-and-ink sketches and advised, "For close geologic observation try drawing the scene and see how much more one learns by looking at it closely enough to draw it!"

After the United States entered World War II, Hunt joined the new Military Geology Unit, among whose responsibilities was the preparation of invasion maps of North Africa and Sicily. One of the most clever, if not the most militarily significant, uses of geology by the unit came in response to the incendiary bombs that the Japanese launched against the West Coast, partly as a psychological weapon but also in the hopes they would land and ignite fires in the forests of the Pacific Northwest. The bombs ascended from an unknown launch site in Japan, traveled east on the jet stream, and fell when instructed to do so by an onboard timer. By studying the minerals and microfossils in the sand in the ballast bags that accompanied the balloons, American scientists narrowed the source of the sand to a small area near Sendai. In early 1945, after photoreconnaissance identified the launch locations, origin of some 9,000 of the balloons, U.S. B-29 heavy bombers obliterated the sites.

After the war, Hunt returned to the USGS to become Regional Geologist for Utah and later chief of the General Geology Branch in Denver, locations that allowed him to pursue his career-long interest in the geology of the Colorado Plateau. His well-earned facility for writing and his dedication to his profession led him to help launch the magazine *GeoTimes*. He left the Survey in 1961 for a series of academic posts at Johns Hopkins, New Mexico State, and the University of Utah. In retirement, he retraced the steps of Gilbert in the Henry Mountains and edited Gilbert's field notebooks, which he had found lying forgotten in the Denver office of the USGS. Hunt wryly noted one passage where, right after Gilbert had written "fresh moccasin tracks," the quality of his notes rapidly deteriorated.

Hunt's first publication on the Grand Canyon appeared immediately after the war. Blackwelder had argued that the Colorado River was a young stream that did not exist until early Pleistocene time. By then, Blackwelder had argued, the essential ingredient of any river, its water, was available from snow melting in the high Rockies. In a 1946 paper, Hunt took the opposite position, arguing that the Colorado River had been a mature drainage system since the early Tertiary. The two models disagreed by many millions of years, yet both authors cited Longwell to support their positions.

Longwell ended up siding more with Blackwelder, but both of them focused only on the stretch of the Colorado just beyond the mouth of the Grand Canyon, where the Muddy Creek Formation crops out. Hunt instead considered the entire river system. In 1956, the Survey published his "Cenozoic Geology of the Colorado Plateau," in which he expanded on his earlier argument that the river is old. Hunt concluded that prior to the general uplift that created the Rocky Mountains (geologists call it the Laramide Orogeny; see "Key Terms and Places") that began in the late Cretaceous, the streams of the Plateau above the Grand Canyon, like those of the Basin and Range province today, remained trapped in their individual basins. But before the uplift had progressed very far, the rejuvenated streams began to cut canyons through the barriers that separated the basins. Hunt's interpretation of the field evidence led him to conclude that the Colorado above the Canyon is at least as old as the Oligocene.

Hunt summed up the problem that confronted the geologists of the 1950s: "The Muddy Creek Formation in Grand Wash trough [just west of the Grand Wash Cliffs, see Figure 8 on page 188] is the immovable object; the irresistible force is the evidence upstream that the Colorado River must have discharged through Grand Canyon before late Tertiary time. Somewhere between the head and foot of the Grand Canyon we lose 10 or 15 million years of river history."

If the entire river is younger than the Muddy Creek Formation, how and when did it create the Grand Canyon? Powell and Dutton had espoused strict antecedence. Though twentieth century geologists recognized antecedence as a possibility elsewhere, none any longer believed it applied to the entire Colorado River; most doubted that it applied to any part of the river. Several had said the Colorado is superposed, but by the mid-twentieth century it was apparent that superposition could not be the whole story either.

Longwell had mentioned one additional possibility: the river might have eroded its way headward, carving out the canyon as it went, finally eating its way across the Kaibab and linking up the lower and upper sections of the Colorado. Strahler agreed that faulting, which dropped the downstream section, would have lowered base level and stimulated the river to greater conquests upstream. But Hunt ridiculed the idea, saying "it would have been a unique and precocious gully that cut headward more than 100 miles across the Grand Canyon section to capture streams east of the Kaibab upwarp."

By the 1960s, things were in a muddle, with irresistible forces colliding with immovable objects and a river with downstream sections that might be younger than the upstream. Geologist Ivo Lucchitta, of whom we will hear more later, put the issue this way: "A stable, integrated drainage system cannot behave in this manner: either some of the dates were wrong, or some of the evidence had been misinterpreted, or the Colorado River has had a much more complex history than formerly recognized." Unfortunately for geologists trying to solve this grandest of geologic puzzles, all three might be true.

By THE EARLY 1960S, MANY still-active geologists had studied the Grand Canyon; hundreds had poured over the Colorado Plateau, particularly during the intensive search for uranium deposits during the late 1940s and 1950s. Each likely had an opinion as to which of Lucchitta's three possibilities was preferable: wrong dates, wrong interpretation, or surprisingly complex river. But all would have agreed that the key questions were: "How old is the Muddy Creek Formation?" and "Before the Basin and Range section existed, where was the upper section of the Colorado River?" It was time to bring the specialists together for the first conference on the history of the Grand Canyon.

In August 1964, under the leadership of Grand Canyon geologist Edwin McKee (1906–1984), the Museum of Northern Arizona in Flagstaff convened a Symposium on the Cenozoic Geology of the Colorado Plateau. The attendees surely knew that in only a few days they would not settle the then nearly-century-old question, but at least they could find out where they agreed and disagreed and chart the direction for future research.

McKee, then regarded as one of the leading geologists of the Canyon, had received an early introduction to its wonders. The leader of his Boy Scout troop in his hometown of Washington, D.C., was Francois Matthes, who in 1903 had made the first topographic maps of the Grand Canyon. In 1927, when McKee was a 21-year-old student at the Naval Academy, Matthes arranged for him to intern with paleontologist John C. Merriam, President of the Carnegie Institution of Washington, who was developing programs for Grand Canyon National Park. Thus began for Ed McKee a lifelong love affair with geology and the Grand Canyon.

In February 1929, after the Park Naturalist tragically drowned while trying to cross the Colorado River, the National Park Service appointed McKee part-time in his place. A few years later, after earning his B.A. in geology from Cornell, McKee returned to the canyon full-time. There he met and began to court biologist Barbara Hastings. Interceding in their romance was the same impediment that had confronted Cardenas and his men in 1540 and everyone since: McKee worked on the South Rim, while his sweetheart worked on the North. A vast chasm separated the two. But to a young man in love, what is a 25-mile hike down into the Grand Canyon and up the other side, only later to have to make the same trip in reverse? His devotion and stamina bore fruit, for on the last day of 1929, the two married. The Park Service had a policy of rotating rangers to another site after ten years. But when his time came to leave, McKee elected to resign and become director of research at the nearby Museum of Northern Arizona. He went on to chair the Geology Department at the University of Arizona in Tucson and then to a position with the USGS. Although he published in several fields, his boyhood interest in the Grand Canyon never left.

By the time of the symposium, although the techniques for measuring rock ages by radioactive parent-daughter pairs were still rudimentary, geochronologists had been able to date several units critical to the history of the Grand Canyon. Because sedimentary rocks seldom contain the radioactive minerals needed for dating, geologists cannot measure their ages directly. But sometimes they can do so indirectly by dating volcanic rocks that interlayer with the sedimentary rocks, and therefore have the same age. Sometimes they can date a volcanic rock below and another above and thus bracket the age of the sedimentary formation that interests them. Fortunately, as Gilbert had noted with regret in the days before the discovery of radioactivity, the Grand Canyon region, especially through the Toroweap section, is rife with basalt flows. By the time of the publication of the Symposium report in 1967, three radiometric dates in particular appeared to constrain the age and history of the Colorado River and the Grand Canyon. Paul Damon of the University of Arizona measured these ages using the decay of potassium to argon.

• Near the top of the Muddy Creek Formation lies a lava flow called the Fortification Basalt. At the time of the symposium, geologists had measured its age at 10.6 ± 1.1 million years. That would have placed the Muddy Creek in the late Pliocene, as Longwell had suspected. But, confusingly, because geologists later shifted the Miocene-Pliocene boundary, we would now say that 10.6 million years is Miocene; see the paragraph beginning at the bottom of page 171.

• At Sandy Point, between the mouth of the Grand Canyon and present-day Lake Mead, a basalt flow with an age of 2.6 ± 0.9 million years rests atop Colorado River gravels and thus must be younger than the gravels. The large error band means that the odds are two out of three that the true age is between 3.5 and 1.7 million years. Thus, the probability is that at some time within that range, the river already flowed west of the mouth of the Grand Canyon at Grand Wash Cliffs (see Figure 8, page 188).

• Farther to the east, in the Toroweap section, volcanoes once spewed lava fountains into the air and sent cascades of molten rock splashing down into the river channel, where they died a hissing death. One particular lava gave an age of 1.2 ± 0.6 million years, again with a large error band. From its relationship with adjacent river deposits, geologists could tell that this particular lava spilled out when the river was only a few tens of feet higher than its present level.

Figure 8. *The Grand Canyon District*

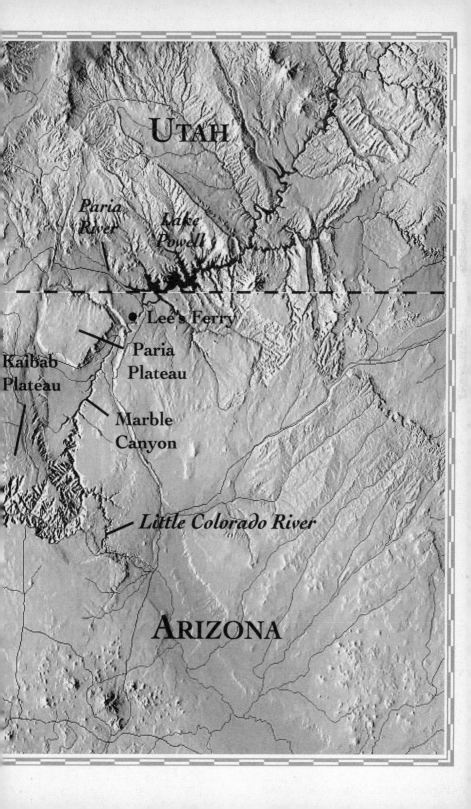

In the early 1960s, scientists were still learning how to use radiometric dating. Naturally, improved methods have caused many of these early dates to be revised. (For instance, a later age measurement dated the Fortification Basalt and the Muddy Creek Formation at 5.9 million years. The date on the Toroweap basalt [the last example discussed previously], even considering the large error, appears to be due to contamination or excess daughter argon, a problem I discuss later.) Nevertheless, the three dates did show the potential of age dating to slice through the Gordian knots of Grand Canyon geology and reveal that the modern Grand Canyon is not fifty or sixty million years old, as the pioneers thought, but only a few million. Though exactly how few, they could not yet say.

The twenty specialists gathered in Flagstaff divided the region into sixteen areas stretching from the lower Colorado River well south of Lake Mead, through the Grand Canyon, and on up to the Hopi and eastern Navajo Reservations. They separated the geologic history of this enormous region into five stages, extending from the end of the Cretaceous to the present. Thus, they had a matrix with geographic region along one axis and time along the other, a useful device for making sure that the conclusions reached about one area jibed with those for the adjacent ones. A final section of the report summarized the theory that had emerged from the conference. I will refer to it as the McKee theory.

As always, the two most important facts were the relatively young age of the Muddy Creek Formation and the antiquity of the Kaibab Uplift (also called an upwarp). The McKee report agreed with Strahler that the arch had formed during the Laramide Orogeny. Thus, for many millions of years, there must have been "at least two different drainage systems, separated by the Kaibab upwarp." This conclusion hearkened back to Longwell's speculation that above the Grand Canyon—east and north of the Kaibab—the river might have flowed south, while Hunt's "precocious gully" approached the Kaibab from the west.

To understand the McKee theory, examine Figure 8. Let your eye follow the Colorado River from Lake Powell past the square-shaped slab of the Paria Plateau and into straight-running Marble Canyon. Ahead,

directly astride the path of the river, awaits the seemingly insurmountable bulwark of the Kaibab Plateau. As the river nears, it first begins to
swing south, even slightly southeast, as if planning to avoid the mountain-lying-down altogether. But just after the Little Colorado River
enters, the main stem swings west to cross the nose of the Kaibab Uplift.
Continuing its swing, the Colorado winds up heading northwest, at
right angles to its direction through Marble Canyon, toward its junction
with the canyon of Kanab Creek. But if one avoids becoming fixated on
present-day topography, a second look reveals an alternate route that the
river might just have taken.

Notice how, right before it begins to change direction at the entry of
the Little Colorado, the main stem Colorado could have avoided bending
simply by continuing on up the valley of the Little Colorado. To be sure,
today that would be impossible, for the Colorado River would have had to
run uphill. But what is now uphill, in a geologic yesterday, may well have
been downhill. Even geologists must remind themselves that the present
is merely one insignificant moment out of hundreds of millions of years.
In the near past, landscapes were different; in the near future, they will
be different again. To understand what happened millions of years ago,
we have to reconstruct the ancient topography of the region. Erosion has
removed more than a mile of Mesozoic sedimentary rock from the
Colorado Plateau, giving a lot of leeway for rivers of the past to have
flowed in different directions than today. In fact, they must have.

Thus, no one can say that as the ancestral Colorado River flowed
south out of Utah, it might not have turned slightly to the southeast and
continued down (now up) the valley of the Little Colorado. If that had
been its path, the Colorado could have joined the ancestral Rio Grande
and flowed on to the Gulf of Mexico. One can pick out a possible path in
Figure 1. Geologists *might* then find deposits of river-laid sediments
along the course of the Little Colorado that would confirm that it once
drained to the south. One sandstone formation along that path does
have cross-beds that suggest southward drainage, but its age is uncertain, so it provides no particular support.

If the Little Colorado once flowed south, what caused it to reverse
direction? And how did the Colorado River west of the Kaibab evolve?

The attendees took their lead from Gilbert, who had shown how a stream could work its way upstream, lengthen its channel, and eventually intersect and capture another drainage. He noted that when the gradient of a stream increases, it speeds up, gains energy, and erodes more rapidly. Think of a stream flowing over a gentle slope and then down a steeper one—a waterfall being the extreme example. Right at the sharp angle where the gentle slope and the steeper one meet—the knickpoint—erosion is greatest. Therefore, the knickpoint, like Niagara Falls, retreats upstream, leaving a deeper channel in its wake. As this happens to every rill and rivulet, systems of streams erode their way upstream, dissect landscapes, and capture their less-energetic cousins. If the base level of a headward-eroding stream should happen to drop, the stream retreats upstream even faster and leaves an even deeper channel behind. As we saw in the last chapter, in 1936 Longwell had speculated that "a stream emptying into the Gulf of California may have extended its valley by headward erosion, starting the cutting of the canyons and finally integrating with drainage from the Rocky Mountains." The same thought occurred to Strahler.

The McKee Report included several sketch maps to illustrate the theory that emerged at the Symposium. In the upper frame of Figure 9, taken from the report, we see the ancestral upper Colorado River running north to south, its headwaters hundreds of miles away in Colorado and Wyoming, its mouth almost as far away, in the Gulf of Mexico. The Kaibab Uplift blocked it from flowing west. There, streams of the Hualapai System drained the land. The Hualapai main stem flowed west and southwest to pool up in a lake in which sediments that will one day become the Willow Springs Formation accumulated. (The Willow Springs Formation, now called Coyote Springs, is not critical to our story.) Basin and Range faulting then began to lower the region to the west, creating the Grand Wash Cliffs, the westernmost of the lines separating D and U. In one or more of the isolated basins, the sediments of the Muddy Creek Formation accumulates.

In the next-younger stage, shown in the lower frame, a vigorous stream has worked its way northward from the Gulf of California (see Figure 15, page 216) into the vicinity of the Muddy Creek sediments

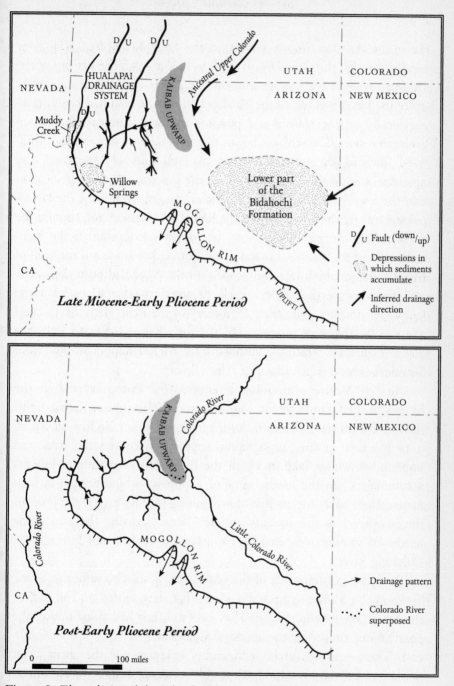

Figure 9. *The evolution of the Colorado River from the McKee Report.*

along the Arizona-Nevada border in the Lake Mead area. The river eroded through the Grand Wash Cliffs to link up with the streams of the Hualapai system. The linkage immediately lowered by a thousand feet or more the base level for all the Hualapai streams, which had been draining locally on the plateau and ponding in lakes, stimulating them to greater erosion. The streams began to cut rapidly downward and headward, incising themselves in their channels and working their way upstream toward the Kaibab Uplift. In the process, they set the location that the lower Grand Canyon would later occupy. Eventually, the Hualapai streams reached and crossed the Kaibab to intersect and capture the main stem Colorado River. Now the previous extension of the main stem, east of the Kaibab, instead of continuing south toward the Gulf of Mexico, did an about-face and headed north. What had been the widest and proudest section found itself demoted to the status of Little Colorado, nothing more than a tributary of the main stem, like a child adopted by the very pirates who had forced its parents to walk the plank. Figure 10, which the authors of the McKee Report adapted from Strahler's second paper, illustrates their theory.

To the McKee symposium attendees, the critical events in the history of the Grand Canyon were structural. The ancient Kaibab uplift had warded off the southward-flowing ancestral Colorado River, keeping it to the east, until a topographic depression blocked the river and impounded a large lake in which the Bidahochi Formation sediments accumulated (see the lower frame of Figure 9), which geologists have named Hopi Lake. Meanwhile, downfaulting to the west greatly lowered the base level of the streams on the Plateau, causing them to erode headward so effectively that they crossed the Kaibab and bisected the Colorado River.

Forty years later, some of the senior geologists who were the Young Turks at the McKee symposium admit that they settled on the McKee theory largely by elimination. They believed they had ruled out all the possible exit routes for the ancestral Colorado River except one: southeast. There was no direct sedimentary evidence for the theory, but neither was there any Muddy Creek-like evidence against it. No one at the meeting felt impelled to speak against the idea. Thus, as so often

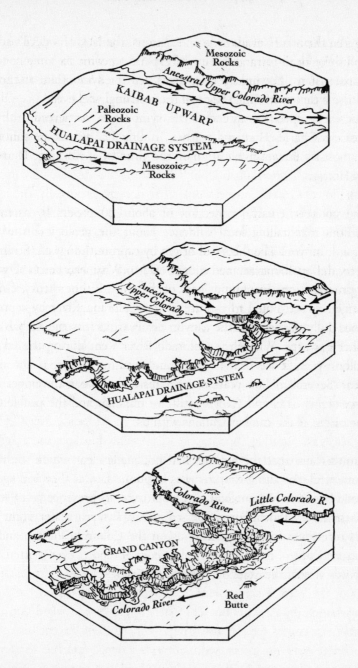

Figure 10. *Evolution of the Upper Grand Canyon, adapted from the McKee Report and Strahler.*

happens in the affairs of humans and nations, the lack of a devil's advocate, plus the need for a gathering of experts to come to some conclusion rather than throwing up their arms, led the symposium attendees to endorse a theory in which none of them strongly believed.

Not every specialist attended the symposium or agreed with its conclusions. Charles Hunt did neither. In only 122 words, he summed up the most significant publication on the Grand Canyon since Dutton's *Tertiary History*:

> The book is a useful collection of about 20 papers by as many authors marshalling local evidence about the geologic history of Grand Canyon. The theory of origin by capture (Longwell, Strahler) is favored rather than antecedence (Dutton), superposition (Davis), or anteposition [Hunt's amalgam of the two]. By the theory, Grand Canyon is attributed to capture of the Colorado River by a precocious gully that eroded the canyon headward across the Kaibab and other upwarps. Where the Colorado River went during the several million years it was in the vicinity and awaiting capture is not made clear. Several million years of river history between the upper and lower ends of Grand Canyon still are missing, and the problem of the origin of the canyon remains with us.

Hunt's claim that the authors had not made clear where the river had gone while "awaiting capture" was not quite fair, as they had speculated that the river had pooled up in Hopi Lake, its presence revealed by the lacustrine Bidahochi Formation sediments. Two years later, Hunt laid out his own ideas in a magnum opus on the Colorado Plateau and its rivers, surely one of most magisterial geological reports ever written. His theory was almost the exact opposite of McKee's.

Chapter 13

Canyon Makers

〜〜〜〜〜〜〜〜〜〜〜〜〜〜〜〜〜〜〜〜〜〜〜〜〜〜〜〜〜〜〜〜〜〜〜〜〜〜

In 1969, two years after the report from the McKee Symposium, the U.S. Geological Survey published one of its prestigious Professional Papers in honor of the 100th anniversary of John Wesley Powell's first journey down the Colorado. Historian Mary C. Rabbitt wrote a biographical sketch of Powell, emphasizing his seminal contributions to the Survey. Ed McKee discussed the stratified rocks of the Grand Canyon. Luna B. Leopold, geologist son of the great pioneer ecologist Aldo Leopold, visited again the question that Gilbert had addressed nearly a century before: why does the Grand Canyon have well over one hundred rapids yet nary a waterfall? Though he used more sophisticated methods, Leopold came to the same conclusion as Gilbert: the Colorado River has had both the time and the energy to eliminate any obstructions in its bed. But the pièce de résistance of the tribute to Powell was Charles Hunt's commanding Geologic History of the Colorado River.

Hunt had worked in the West for years; no one was better qualified to attempt to explain the overall history of the Colorado Plateau and its rivers. He began his survey high in the Colorado Rockies, at the headwaters of the Grand, and on the Green above the Uinta Range. He traced these rivers and their major tributaries downstream toward that obstacle to rivers and geological theories: the Kaibab Plateau, explaining each of the major geologic features encountered on the way. His analysis and years of fieldwork led Hunt to interpret the Colorado above the Kaibab as dating back to the early Oligocene, some twenty-five or thirty million years ago. But Longwell and others had interpreted the evidence west of the Kaibab to show that the river did not exit from its present mouth at Grand Wash Cliffs until after the Muddy Creek sediments accumulated, at the time thought to be about ten million years ago. This led Hunt to write that "Somewhere between the head and foot of the Grand Canyon we lose 10 or 15 million years of river history." This gap produced what he called the "Grand Canyon—Grand Problem." Remember that, to a geologist, calling something a problem is a compliment.

Hunt had already rejected peremptorily both headward erosion and the theory that the river had flowed southeast to join the Rio Grande. He also thought the Colorado had always been a single, integrated stream, rather than segments of disparate rivers stitched together. Therefore, if one section of river is old, so must it all be. Thus, Hunt concluded that the Colorado River had flowed west of the Kaibab at the time the Muddy Creek sediments accumulated, yet those very sediments show that the river did not then exit at the present mouth of the Grand Canyon. Where had it gone? A major river ought to have left some evidence of its former passage down this alternate route, such as an abandoned channel and ancient river sediments. Hunt's analysis required him to find that long-deserted route.

His sketch map of the region southwest of the mouth of the Grand Canyon, Figure 11, shows the location of Peach Springs Canyon. It aligns with the Hurricane Fault, which lies along the east flank of the Shivwits Plateau (shown in cross-section in Figure 3, page 48) and runs through the hamlets of Peach Springs and Truxton. (You can also see the canyon in Figure 9, page 193, and unlabeled in Figure 1.) Hunt claimed that during the early Miocene, Peach Springs Canyon had sloped toward the south and southwest, providing the exit route off the Plateau that he sought.

Around the time Hunt was preparing his masterpiece, Richard Young, a graduate student at Washington University, was conducting the fieldwork for his Ph.D. thesis, entitled "Cenozoic Geology along the Edge of the Colorado Plateau in NW Arizona." His area was the Huala-pai Plateau, shown on Hunt's sketch map and in Figure 8, page 188.

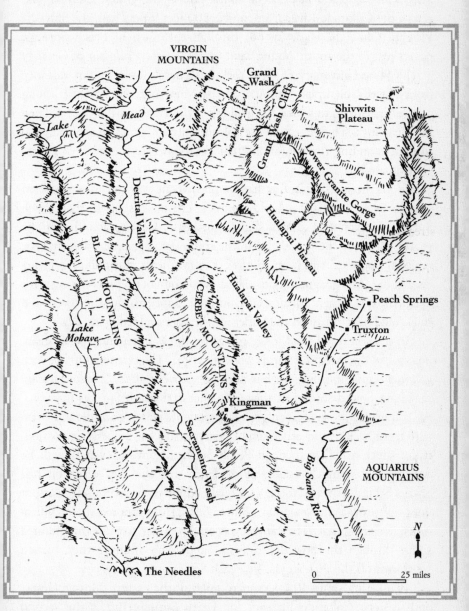

Figure 11. *Hunt avoids the immovable.*

This remote upland lies on the south side of the Grand Canyon, just across the Colorado River from, and 1,500 feet lower than, the Shivwits Plateau. Old U.S. Highway 66, through the town of Peach Springs, passes just to the south. Young went on to become Professor of Geology at The State University of New York at Geneseo, only a short distance from Powell's birthplace at Mt. Morris, New York. He has had a long and varied career, in which the geology of the Hualapai Plateau has been his life's work. The plateau is critical not only to Hunt's theory, but also to understanding the conditions that existed before the modern Grand Canyon appeared.

In his thesis work, Young had found that the way the ancient stream gravels overlapped in the old bedrock channels showed that Peach Springs Canyon had always sloped northeast. He also found that the source rocks for the gravels lay to the southwest, again showing that the regional slope had run from southwest to northeast, downhill. Nevertheless, in his 1969 magnum opus, Hunt cited Young's thesis as the basis for his claim that Peach Springs Canyon once sloped to the southwest and thus could have provided the outlet for the ancestral Colorado River.

In the section of the McKee report on the western Hualapai Plateau, Young had drawn on his thesis work to write, "The relationship between channel segments suggests, but does not prove, that a Cenozoic stream may have flowed southwest down the ancestral Peach Springs Canyon." As Young tells it, as a green graduate student, he felt considerable pressure from McKee and other senior geologists to leave open the possibility of southwest flow through Peach Springs Canyon, even though he had actually found considerable evidence that the flow direction had always been northeast. Young had qualified his report with "suggests but does not prove" and "may have flowed," but Hunt ignored the qualifications, declaring that "The canyon was cut by earlier streams that drained south through the now dry canyon," and "The drainage then turned southwest off the Plateau via the Peach Springs dry valley."

Blanketing some 40,000 square miles of the eastern Mojave Desert and western Colorado Plateau, the Peach Springs Tuff is the second largest of its kind in the world. Erosion has turned a section of the tuff into the "Needles" on the Colorado River, for which the gateway town in

California is named. Tuffs form when volcanoes blast red-hot, molten magma high into the air. As the micro-droplets settle back to earth, they weld together into solid rock—the tuff. These explosions of fiery rain, as the few survivors at Pompeii and Martinique could have testified, are the most dangerous of all volcanic eruptions. A rock that formed instantaneously and that is as widespread as the Peach Springs Tuff would provide geologists with a most useful marker bed, one they could use to reconstruct the post-tuff geological history of the desert southwest, if only they could date the tuff.

This may be a good place to recall that fossils provide only the relative age of a rock. Igneous and metamorphic rocks contain no fossils; many sedimentary rocks such as sandstones and conglomerates have few if any useful ones. To allow geology to progress much beyond where it found itself by the mid-twentieth century, its practitioners had to find a way to measure the absolute ages—not just the relative ages—of many more rocks, especially the igneous ones. They found their solution in the inexorability of radioactive decay. The measuring of absolute rock ages was as important as the discovery of standard candles in astronomy, which enables astronomers to tell the distance between galaxies. One could even say that the discovery of "standard candles" in geology—the ability to measure absolute ages—was as important a twentieth-century breakthrough as the discovery of DNA in biology, because without both of these pieces of scientific knowledge, evolution would make little sense. Certainly without absolute age dating, our understanding of the Grand Canyon would have stalled decades ago.

Peach Springs Canyon not only had to be in the right place and slope in the right direction for Hunt's theory, it had to be there at the right time: before the Muddy Creek sediments accumulated. The Peach Springs Tuff interlayers with the sedimentary rocks that fill the canyon, so that the canyon must have existed and been available as a possible exit route for the Colorado before the tuff fell. The invention of potassium-argon dating allowed geologists to establish the tuff's age at 18 million years, making it significantly older than the Muddy Creek Formation, which at the time appeared to be no older than about 10 million years. So far, so good for Hunt's theory.

But the tuff produced as many problems for Hunt as it solved. The tuff and the gravels interbedded with it had so completely blocked Peach Springs Canyon that it could no longer have served as the exit route for the ancestral Colorado. Yet the more recent date on the Muddy Creek Formation, mentioned above, showed that the river did not begin to exit through the Grand Wash Cliffs until after five or six million years ago. Where was the ancestral Colorado between the time the Peach Springs Tuff fell 18 million yeFars ago and blocked the river's putative path down Peach Springs Canyon, and the time at which it breached the Grand Wash Cliffs, five or six million years ago? Hunt had closed one gap of "10 to 15 million years of river history" only to create another of roughly the same length. In order to close this second gap of his own making, Hunt had to invent a theory of subterranean conduits that even he admitted was "outrageous."

The three sketch maps in Figure 12 illustrate Hunt's idea. The oldest, in the upper left, shows that according to Hunt's interpretation, by the Oligocene and early Miocene, some 25 million years ago, the Colorado Plateau already had an extensive drainage system (dashed lines) whose streams lay close to their modern locations (gray solid lines).

Hunt thought that the Little Colorado River, rather than the main stem, had first cut its way across the Kaibab Uplift, called the Kaibab-Coconino Uplift on the sketch, possibly a little south of the present course of the river. West of the Kaibab, the river left the plateau via the dry canyon at Peach Springs, which he interpreted as having then sloped to the southwest. From there the river flowed past Kingman and on to the Gulf of California. Hunt's difficulty was to close the multi-million year gap—to explain how the ancestral Colorado River got from its exit route through Peach Springs Canyon to its present location in the Grand Canyon without leaving river sediments on the Hualapai Plateau in between to mark its trail in between—sediments that geologists have never found. Hunt had to engage in some remarkable geologic contortions.

The middle frame of the sketch map shows the disposition of the rivers and structures after eighteen million years ago. By now, the streams above the Kaibab have moved closer to their present courses;

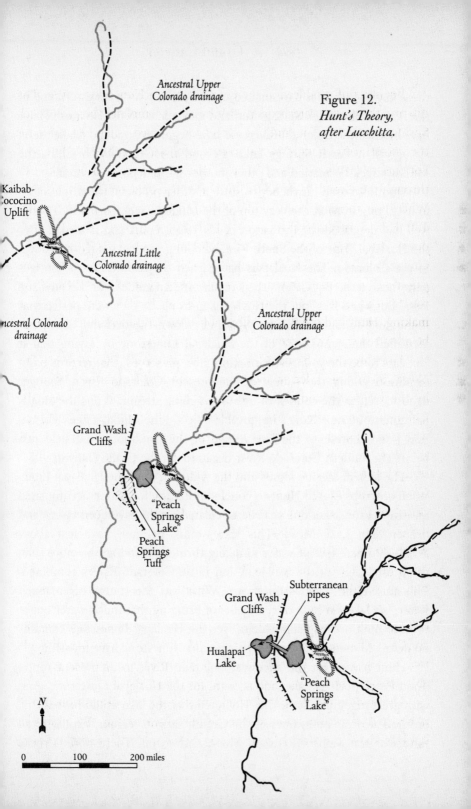

Figure 12.
*Hunt's Theory,
after Lucchitta.*

Ancestral Upper
Colorado drainage

Kaibab-
Cococino
Uplift

Ancestral Little
Colorado drainage

Ancestral Colorado
drainage

Ancestral Upper
Colorado drainage

Grand Wash
Cliffs

"Peach
Springs
Lake"

Peach
Springs
Tuff

Subterranean
pipes

Grand Wash
Cliffs

Hualapai
Lake

"Peach
Springs
Lake"

N

0   100   200 miles

those cutting across it have incised more deeply. Near the western end of the Plateau, the Peach Springs Tuff and movement on the Hurricane fault have blocked the Colorado River and have begun to impound a lake there; let us call it Peach Springs Lake. Beyond the Grand Wash Cliffs, the Muddy Creek Formation and other interior basin deposits accumulate.

Hunt also used his theory to solve another problem: the origin of an unusual limestone at the very top of the Muddy Creek Formation. Longwell had described the formation in his 1936 report and named it after the Hualapai Tribe of the South Rim, who long ago had guided Ives down to the Colorado. The Hualapais had earned the tribute, for as the Ives party neared the Big Cañon, their regular Indian guides had "become at a loss." But when Ives "put the Hualapais in front . . . they were perfectly at home, and conducted [his party] rapidly down the declivity." Longwell identified the significance of the Hualapai Limestone by saying that it "appears to be the youngest formation, now preserved, that records basin conditions along the course of the present Colorado River. Younger deposits reflect the control of a through-going stream." Thus the Hualapai Limestone caps Hunt's immovable object—the Muddy Creek Formation—and represents the last vestige of the conditions that held just before the modern Colorado River began to cut the Grand Canyon.

The lack of marine fossils and the wide extent of the Hualapai Limestone outcrops caused Hunt to conclude that the formation accumulated in a freshwater lake, one that he calculated would have been larger and deeper than Lake Mead. This lake would, he said, have lost about 2,750,000 acre-feet of water annually through evaporation, more than the present flow of the San Juan and Little Colorado Rivers combined. This raised the obvious question: "What was the source of so much water?" A lake fed by a river capable of bringing that much water ought to have built a conspicuous delta, yet the Hualapai Limestone contains no delta sediments. How could this peculiar limestone have formed?

Hunt had to get the ancestral Colorado River to its present route, drain Peach Springs Lake, and account for the Hualapai Limestone. And with great ingenuity, so he did. He noted that the lake would have rested on the Paleozoic limestones of the Grand Canyon region, familiar to all rafters as the Kaibab, Redwall, Muav, and so on. These porous rocks would have allowed the lake waters to seep down and leak out through underground channels, or, as Hunt referred to them, pipes. The carbonate-rich underground waters would have flowed downslope to emerge west of the Grand Wash Cliffs as springs feeding Hualapai Lake. The sediment carried by the ancestral Colorado, now gone underground, would have been trapped in a maze of subterranean channels, leaving the emerging springwater clear, and explaining why there are no delta deposits intermingled with the Hualapai Limestone. The limey water precipitated the Hualapai Limestone on top of the Muddy Creek sediments. Thus, Hunt envisioned not one but two lakes, the upstream Peach Springs Lake feeding the lower Hualapai Lake subterraneously. Later and back upstream, the Green and Colorado merged with the San Juan and Little Colorado to produce a greatly increased volume of river water. The emboldened stream cut a new channel north of Peach Springs Canyon, drained the lakes, and in the process began to carve the Grand Canyon. Whew!

Four decades after Hunt wrote, age dating and better understanding of regional geology have negated some of his key assumptions. Young's evidence that Peach Springs Canyon flowed northeast grew stronger with the years; it alone is enough to falsify the core of Hunt's theory. If the ancestral Colorado that Hunt envisioned was anything like the muddy river that Powell rode, it is hard to understand what would have happened to the vast amount of sediment the river carried as it went underground. It would have made a delta at the point where the river sank out of sight, clogged the subterranean pipes, or be present beyond the Grand Wash Cliffs. Still, where so much erosion has gone on, one cannot say that it could not have removed such delta sediments.

Although he defended his idea vigorously again in 1974, Hunt surely knew that if his predecessors, in a century of trying, had not succeeded in explaining the Grand Canyon, he might well not have done so either. He certainly knew that there was no direct evidence for Peach Springs Lake—it was an inference, an educated geologic guess. Perhaps he put forth his "outrageous" theory partly as a straw man, to show that the McKee theory had at least one serious rival (from the Latin *rivalis*: one living on the opposite bank of a stream from another).

He intended to write another treatise on the history of the Colorado River, but, sadly for those who would try to understand the river and its canyons, never did. Here is how Hunt ended his magnum opus:

> As a body of water, the Colorado River is small. Its flow is only 5 to 10 percent of that of other great rivers in the United States, such as the Columbia, Snake, Missouri, Ohio, and St. Lawrence. But the Colorado River is a major geographic force in the American southwest. The river crosses the arid lands that need its water, some of the arid lands about which Powell concerned himself. Because of its physiographic setting and geologic history, the Colorado River basin is spectacularly scenic. Nearly half our national parks and monuments are within the Basin. Even after 100 years, however, the explanation of this landscape still defies us.

McKee and colleagues routed the ancient Colorado River southeast down the valley of the Little Colorado and on to the Gulf of Mexico; Hunt continued it southwest through Peach Springs Canyon to the Gulf of California. But both authors had arrived at their theories partly by elimination and partly by inference: no direct evidence ever turned up to support either. Nothing is wrong with an educated guess; indeed, in trying to trace the history of a river, one has to make them. But if after a while no evidence turns up to support a profession's educated guesses, it is time to make others. If the river went neither southeast nor southwest, it could only have flowed upstream, the way it came (a seemingly unlikely proposition) or run northwest. Geologist Ivo Lucchitta had attended the McKee Symposium and signed on to its conclusions, but the absence of evidence for the path down (now up) the Little Colorado caused him to change his mind. In 1984, he proposed that the Colorado River had crossed the Kaibab and headed northwest into Utah and Nevada.

Lucchitta was born in central Europe just before World War II, received a bachelor's degree in classics in Rome, and in 1956 emigrated to the United States. He then switched fields, earning a bachelor's degree in geology from the California Institute of Technology and, in 1966, a Ph.D. from Pennsylvania State University. Ed McKee persuaded Lucchitta, self-described as "enthusiastic and foolish" at the time, to do his doctoral thesis on a Grand Canyon problem; its title was "Cenozoic Geology of the upper Lake Mead Area adjacent to the Grand Wash Cliffs, Arizona." Lucchitta began his career with the USGS in Flagstaff, Arizona, but instead of continuing to work on his nearby thesis area and the Grand Canyon, he found that his Survey assignment took him, figuratively speaking, 270,000 miles away. But he never lost the fascination that the Grand Canyon has held for so many geologists; over a 35-year career, he found time to write a series of provocative papers on its origin.

As we have seen, Young's work ruled out Hunt's theory that Peach Springs Canyon had served as the exit route for a southwest-flowing ancestral Colorado River. Lucchitta believed that Hunt was also wrong about the Hualapai Limestone. He said the unit does not center on the mouth of the Grand Canyon, but is widespread, requiring an even larger lake than Hunt envisioned, or perhaps several lakes and thus an even more outrageous set of subterranean pipes to fill them. Lucchitta viewed the limestone as merely the uppermost of a typical set of interior basin deposits. Finally, Lucchitta thought that Hunt's end run around the Muddy Creek Formation, even if the river had flowed southwest off the plateau, did not really solve the Muddy Creek problem, for "The hypothetical river downstream from Peach Springs Canyon towards the sea is just as plugged up by deposits of interior drainage as the area near the mouth of the Canyon."

Having rejected all the extant theories, it was time for Lucchitta to propose his own. In 1988 the Museum of Northern Arizona published *Canyon Maker*, his most complete, and beautifully illustrated, version. His theory has changed but little over the years; the most recent refinement appeared in 2003 in the second edition of *Grand Canyon Geology*.

Lucchitta proposed that after crossing the Kaibab, the Colorado had taken one of the two paths no one had yet suggested: north and northwest. Lucchitta presented several key pieces of evidence showing that the regional drainage before the Grand Canyon existed had been to the north. First, scattered here and there on the Hualapai Plateau, Young had found ancient, abandoned northeast-trending stream channels— like Peach Springs Canyon—that contain deposits of "rim gravels." These coarse, stream-deposited gravels include rounded pebbles of igneous and metamorphic rocks found only far to the south, in the Mogollon Highlands, or Rim, of Central Arizona, the southern edge of the Colorado Plateau (see Figure 9, for example). Only streams flowing from south to north could have brought the gravels from the southern rim of the Plateau all the way north to the Hualapai Plateau. The second piece of evidence is the 18-million-year-old Peach Springs Tuff, whose age Hunt had used to support his theory. Young had found that the fluid tuff had surged down a channel trending north and northeast, opposite the direction Hunt required.

Lucchitta's third piece of evidence is the drainage direction of the two large streams that enter the Grand Canyon from the south: the Little Colorado River and Cataract Creek, which is called Havasu Creek in its lower section. Except for these two (and Kanab Creek, which enters from the north along the west side of the Kaibab Plateau; Figure 8 shows all three), the tributaries to the Colorado are "short, steep, immature, and clearly related to the cutting of the Grand Canyon. The streams are much too young to be of help in deciphering the pre-canyon drainage patterns." Cataract Creek, and the Little Colorado especially, have long, mature valleys; they likely predate the Grand Canyon and indicate that the drainage then was to the north or northwest, following the overall dip of the rocks.

But according to Lucchitta, even a sizable stream like the Little Colorado is too small to have been the true master stream of such a large section of the Colorado Plateau; instead, both the Little Colorado and Cataract Creek must have been tributary to a larger stream—the ancestral Colorado River. This main stem must have flowed down the same regional slope and in the same general direction as its main tributaries: north and northwest. All this evidence convinced Lucchitta that the direction of the regional drainage had been north, not southeast as in the McKee theory, or southwest as in Hunt's.

If the ancestral Colorado had flowed south through Utah to the east

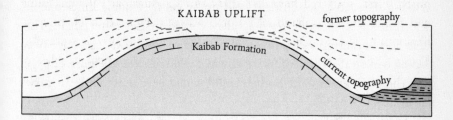

Figure 13. *Lucchitta's cross-section of the ancestral Kaibab Uplift.*

side of the Kaibab Uplift, but after crossing over it had run north and northwest, Lucchitta, like all before him, faced the problem of how to get the river across the giant arch. His reconstruction of the rocks that once existed above the Kaibab—the same Mesozoic strata that Walcott thought had pressed down and kept a flexure from becoming a fault—convinced Lucchitta that the Kaibab problem was not so difficult after all. He thought that the ancestral Colorado River, flowing across the then-existing Mesozoic rocks, thousands of feet above the present land surface, had found the Kaibab region to be a topographic low—a strike valley carved in softer rock, surrounded by cliffs of more resistant ones. The cross-section diagram—an imaginary vertical slice down through the surface—shown as Figure 13 illustrates the current topography and Lucchitta's reconstruction of the former topography. This is a good place to remember that rock resistance matters much more to topography than geologic structure. Instead of a high plateau, the river may have found a broad valley, as shown in the center of the cross section.

One of the most common features of the Colorado Plateau and the Grand Canyon is the "hard-soft couplet," in which a layer of resistant rock lies above a weak one. In the Grand Canyon itself, the resistant sandstones and limestones form the vertical cliffs while the softer shales makes up the sloping sections in between, as shown in Figure 16, page 260, and in any photograph of the Grand Canyon. Indeed, it is this juxtaposition of colorful rocks of varying resistance to arid erosion, with the resulting alteration of cliffs and slopes that gives the Colorado Plateau its unique character. On the plateau above the canyons, the couplets show up as long ridges held up by gently dipping limestone or sandstone beds, looming over a valley carved in the softer shale. The ridge and bordering valley follow the strike of the rocks, the trace of a sloping bed on the surface. As we saw in the discussion of Figure 7, page 177, the rocks around the Kaibab Uplift would have outcropped in curving, concentric bands.

Assembling all this evidence, Lucchitta envisioned the Colorado River coming down onto the Kaibab from the north, then sweeping around it in a giant, arcuate strike valley—a Racetrack of the Titans—and heading off to the north and northwest. Indeed, the Grand Canyon today makes just such a bend, but then at Kanab Creek turns southwest again.

If Lucchitta is right, we might find deposits of old Colorado River sediment north and west of the Grand Canyon. He does say that "gravels of probable river origin [occur] in the area of the Kanab, Uinkaret, and Shivwits plateaus," but this is not much to go on. Geologist George Billingsley mapped these same plateaus without finding outcrops of confirmed river gravel. As with the McKee and Hunt theories, the key evidence that would support Lucchitta's idea has yet to appear, though it still could.

BY THE TIME LUCCHITTA wrote in the 1980s, and in spite of Hunt's ridicule of the idea, he and other geologists had accepted what we might call the lower half of the McKee theory: that a stream had eroded its way headward, breached the Grand Wash Cliffs, and proceeded upstream to intersect the Colorado River near the Kaibab Uplift. The process could have left in its wake the modern Grand Canyon. Can we tell when this might have happened? All the way back to Blackwelder and Longwell, geologists had believed that the presence of the interior basin deposits of the Muddy Creek Formation near the mouth of the Grand Canyon meant that a major river could not have flowed through the area until later. This was a reasonable conclusion, but, like so much else in the Grand Canyon, unproven. It is easy to imagine that an older river might have dried up, or run underground, after which intermittent streams like those in today's Mojave Desert deposited the Muddy Creek sediments. Later, when more water again became available, the same river, or another, reopened the old channel and laid down river gravels on top of the Muddy Creek Formation. But even if it did have an older ancestor, the Grand Canyon that we see today—the modern Grand Canyon—can be no older than the Muddy Creek Formation.

The early age measurements on Grand Canyon volcanic rocks used a technique that depends on the radioactive decay of potassium to argon. Over the last quarter of the twentieth century, geologists had come to replace that technique with a more accurate one called the argon-argon method, which I will discuss in more detail later. The highest, and therefore youngest, of the rocks of the Muddy Creek Formation is Longwell's Hualapai Limestone. Its age would set the maximum for the emergence of the Colorado River west of the Grand Wash Cliffs, and therefore the western Grand Canyon itself, but geologists cannot date limestones directly. Fortunately, as the Hualapai Limestone accumulated in a freshwater lake, a volcano blasted up a cloud of fiery ash, which settled as a tuff. This eruption happened millions of years after the Peach Springs Tuff fell and should not be confused with it. In 1998, geologists used the argon-argon method to date the Muddy Creek Tuff at six million years, revealing that the older age measurement described earlier was erroneously high. The more accurate modern measurement allows us to conclude that the modern Grand Canyon is younger than six million years.

Sediments deposited by the Colorado River would set the minimum age of the Grand Canyon, if geologists could find and date those sediments. Because erosion has removed any that formed, deposits that date back to the youth of the Colorado River are absent in the western Grand Canyon itself. Instead, geologists had to look downstream, south of modern Lake Mead, where the Colorado flows in a more or less normal river valley and where its deposits might still be around. There, structural basins contain a sedimentary rock called the Bouse Formation. It comprises silt, sand, and gravel typical of delta or estuary deposits, as well as interbedded volcanic rocks. The Bouse lies under and fingers up into recognizable Colorado River gravel above; it lies on top of an interior basin deposit that contains "extremely sparse material of probably Colorado Plateau derivation." Thus, the Bouse Formation marks the first appearance of the Colorado River below the Grand Canyon. Fortunately, interbedded with the sedimentary layers of the Bouse Formation is yet another volcanic tuff, this one with an argon-argon age of five million years. By that time, the Colorado River had already begun to cut the Grand Canyon.

Large rivers not only erode headward, they lengthen their channels downstream. As a river runs out of energy near its mouth, it begins to drop its load of debris and build a delta out into the body of water it enters. The river then flows over its delta, extending the sediment wedge farther and farther downstream, sometimes creating a bird's-foot delta pattern of distributaries as in the Nile and Mississippi. The Bouse Formation appears to be the first delta deposit formed when the Colorado River began to enter the Gulf of California. The Imperial Formation, which is just younger and lies just south of the Bouse deposits, represents the downstream progression of the delta. The Imperial Formation contains fossils otherwise found only in a shale high upstream on the Colorado Plateau, again showing that by the time the river deposited the Bouse and Imperial formations, some five million years ago, the upper and lower courses of the Colorado had joined, and the cutting of the Grand Canyon had begun. But there is another, independent way to date the initial appearance of the Grand Canyon. As noted several times earlier, the character of the Green and Colorado rivers changes markedly and repeatedly downstream. None of the changes is more dramatic than the contrast between the western Grand Canyon—incised nearly a mile deep in the plateau—and the river channel from Lake Mead south to the Gulf of California, as Figure 1 plainly shows. But why does the lower Colorado section exist in the first place, and what can it tell us about the age of the Grand Canyon upstream?

The destination of the modern Colorado River is the Gulf of California, shown in Figure 15. Before it existed, the rivers that drained the Colorado Plateau must have emptied somewhere else, either into a still-larger river or into some long-vanished sea or lake. The section of the Colorado River from Lake Mead downstream can be no older than its modern terminus, the Gulf of California.

To assert that "nothing in geology makes sense except in the light of plate tectonics" is no more of an exaggeration than biologist Theodosius Dobzhansky's parallel remark about the role of evolution in biology. Indeed, plate movement created the Colorado Plateau, the Basin and Range Province, the San Andreas Fault, and each of the major features of western North America, including the Rocky Mountains and the Gulf of California. If we know enough about the history of plate activity in western North America and the eastern Pacific, we can date the time of origin of the Gulf of California and therefore derive an independent estimate of the maximum age of the lower Colorado River. If the opening of the Gulf created a knickpoint that traveled upstream and breached the Grand Wash Cliffs from below, as Lucchitta and others believe, the age of the Gulf also sets the maximum age of the modern Grand Canyon itself.

The giant, thin slabs that segment the outer 100 kilometers or so of the earth's upper mantle and crust can move in three directions relative to each other. Two plates can collide; if each carries a continent piggyback, a high mountain range results, as in the Himalayas. If one or neither of a pair of converging plates carries a continent, one plate slides underneath the other in a subduction zone, as in the Sumatran Trench today. Two plates can pull apart, allowing a mid-oceanic ridge and a new ocean basin to arise between them, as in the Atlantic Basin. Finally, two plates can slide sideways past one another, as in the San Andreas Fault. Because these relationships come down to the geometry of thin slabs on the surface of a sphere, geologists can work backwards from present plate motion to deduce past motion.

As shown in Figure 14, some 30 million years ago the Farallon Plate lay between the American and Pacific Plates. The two converged along a subduction zone that gradually consumed the Farallon Plate. By about 20 million years ago, it had vanished, leaving behind two smaller remnants: the Juan de Fuca and Cocos Plates. The Farallon Plate eventually traveled east for 1,500 kilometers, so far underneath North America that it caused the uplift of the Rocky Mountains.

Plate tectonic theory shows that when a subduction zone meets a spreading center such as a mid-ocean ridge, which geologists call a divergent plate boundary, a transform fault develops to accommodate their different vectors of motion. When the spreading center on the leading edge of the Pacific Plate struck the subduction zone between it and the American Plate, shown in the second frame from the top in Figure 14, the San Andreas Fault resulted. The shearing motion between the two giant plates on either side of the San Andreas Fault later caused a piece of the mainland—the American Plate—to break off and weld to the Pacific Plate. This left the Baja California part of the Pacific Plate, and the rest of Mexico part of the American Plate. Since then, Baja California has moved some 300 kilometers to the northwest. Some 70 million years or so from now, Baja will disappear down the Aleutian Trench.

The proto-Gulf of California extended hundreds of miles farther to the north than it does today, into the structurally identical Salton Trough on the California-Mexico border (see Figure 15, page 216). The Colorado River has since deposited many thousands of feet of river sediment like the Bouse and Imperial Formations, enough to fill the Salton Trough and create the Colorado River Delta. Deep wells there have penetrated over 18,000 feet of sediment without hitting bedrock. Geologist Robert Scarborough estimates that 50,000 cubic miles of sediment may lie buried under the Gulf of California. Cut his estimate in half and you still have dramatic evidence of the ability of the streams of the Colorado Plateau to produce and transport sediment.

The best estimate from plate tectonics is that the Gulf of California opened between four and five million years ago. This range conforms to the ages of the initiation of the Grand Canyon obtained by a completely different method—argon-argon dating of the Hualapai Limestone and the Bouse Formation. We know from the age of the Muddy Creek Formation that the modern Grand Canyon is not older than six million years. Thus, geologists have a consistent set of answers to the question of when the modern Grand Canyon began to appear, if not yet why.

Powell, Dutton, and the other pioneers had thought, logically enough, that such a deep canyon must have taken a correspondingly long time to form. That belief, together with their ideas about consequent streams and the evidence of a shrinking Eocene sea, caused them to date the origin of the Grand Canyon to a time some fifty million

30 million years before present

AMERICAN PLATE
• Los Angeles

FARALLON   PLATE

PACIFIC
PLATE

20 million years

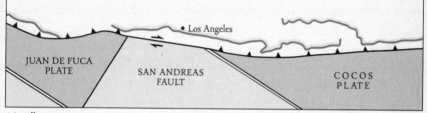

• Los Angeles

JUAN DE FUCA
PLATE

SAN ANDREAS
FAULT

COCOS
PLATE

10 million years

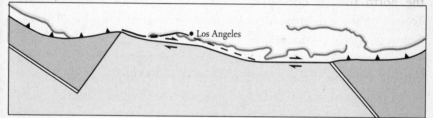

• Los Angeles

5 million years

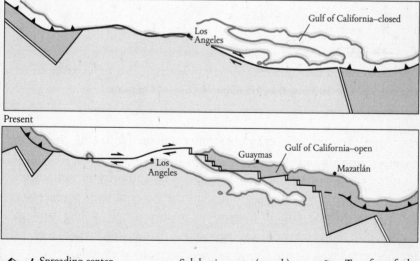

Gulf of California–closed

Los
Angeles

Present

Gulf of California–open

Guaymas

Los
Angeles

Mazatlán

⟨symbol⟩ Spreading center,
divergent boundary

⟨symbol⟩ Subduction zone (trench),
convergent boundary

⟨symbol⟩ Transform fault:
the San Andreas

Figure 14. *Plate tectonics of southwest United States and Mexico.*

Figure 15. *The American Southwest and the Gulf of California.*

years ago. Yet modern evidence from radiometric dating and plate tectonics appears to constrain the age of the western Grand Canyon to no more than four to five million years. The authority of radiometric dating caused many geologists to discard the views of their predecessors and to regard the Grand Canyon as a surprisingly young geologic feature. But not everyone was persuaded. Some saw evidence of yet another paradox: a Grand Canyon that is both old *and* young.

Chapter 14

Lazarus and Lakes

~~~~~~

Now and again in studying the history of science, one can identify a particular meeting at which scientists finally rejected an old theory and adopted its replacement. Occasionally, one can pin down what Thomas Kuhn called a paradigm shift to the few days of a single meeting. Of course, most meetings, like most human affairs, fail to produce a breakthrough. At least one geology conference gave rise to such a strong, but wrong, consensus that it set the field back by decades. In 1926, the American Association of Petroleum Geologists met to evaluate Wegener's theory of continental drift. In a classic example of groupthink, almost to a person the attendees excoriated the theory and disparaged Wegener's methods and motives. Forty-one years later, those attending the annual meeting of the American Geophysical Union overwhelmingly endorsed the modern version of Wegener's theory—plate tectonics—and ratified the most important paradigm shift in the modern history of geology.

The attendees at the 1964 Conference on the Grand Canyon, led by Ed McKee, agreed on a bold new hypothesis: headward-eroding Hualapai streams had bisected the ancestral, southeast-flowing Colorado River. The capture reversed the drainage on the section of the Colorado that had flowed on to New Mexico, demoting that section to the status of north-flowing Little Colorado River. The addition of water from above the Kaibab to the Hualapai main stem greatly increased its volume, in effect turning it into the Colorado River and allowing it to quickly carve the Grand Canyon. But in the years after the conference, no evidence ever turned up to support the half of the McKee theory that applied to the Colorado River above the Kaibab. It died aborning.

Charles Hunt gave the McKee theory short shrift, continuing to say that the entire Grand Canyon, from top to bottom, is old. But Young had already showed that the drainage in Peach Springs Canyon had flowed opposite to Hunt's claim. Ivo Lucchitta's theory differed from McKee's for the region above the Kaibab and from Hunt's for the region below it. Lucchitta had envisioned the ancestral Colorado wheeling around the Kaibab and heading off to Utah and Nevada, but once again, no direct evidence supports the idea. McKee, Hunt, and Lucchitta were hundreds of miles, and millions of years apart, with no concrete support for any of their theories. That the best geologists, who collectively had put in scores of years of work on the Grand Canyon, could not find the evidence to confirm their theories tells us much about the difficulty of trying to find long-vanished rivers.

Four decades after the 1964 conference, McKee and Hunt were gone, but others had stepped forward to take their places. Scores of geologists had put in years of work on the Grand Canyon—the online Bibliography of the Grand Canyon and Lower Colorado River listed well over 20,000 entries. Plate tectonics had arrived to effect a scientific revolution and reveal that the Gulf of California, the object of today's lower Colorado River, had opened only four to five million years ago. So much new information had become available in the decades since the first conference that the leading Grand Canyon geologists decided that it was past time for a second get-together. In June 2000, seventy-seven specialists convened at Grand Canyon National Park for a weeklong symposium of talks and field trips.

Though the attendees made dozens of presentations, none endorsed either Hunt's theory or McKee's. In spite of the great respect owed these two outstanding geologists, the passage of decades without supporting evidence, and accumulating negative evidence, relegated their two theories to the history of geology. Yet before continuing on, we must acknowledge that by stimulating research from proponents and opponents alike, Hunt and McKee made lasting contributions to Grand Canyon geology. In science, being wrong is no vice; it is both inevitable and productive. Scientific progress rests on a scaffold of theories, most of them eventually found wanting.

The conveners of the June 2000 Symposium had planned to get everyone together at the end to work out a consensus opinion. But so many more attended than expected that after the paper presentations, the attendees had no time for the planned meeting of minds. Perhaps this was just as well, for no consensus had emerged. Instead, the Symposium report comprises thirty-three chapters, each covering a different aspect of Grand Canyon geology. Editor Dick Young closed his introduction by posing eleven critical issues that still challenge Grand Canyon geologists. One wonders what Powell and Dutton would have made of Young's list. I like to think that although at first they would have been dismayed, on reflection they would have been immensely proud of their profession.

Though one would not call it a consensus, several authors did support the seemingly paradoxical idea that the topography of the Grand Canyon district is both old and young. In the preceding chapter I reviewed the evidence that the Colorado River carved the Grand Canyon in less than five or six million years. No one at the symposium refuted that evidence—indeed, they presented a good deal more in support—but some pointed to other facts, some of them long recognized, that collectively supported the notion that the canyon might be older. How were geologists to reconcile the seemingly irreconcilable evidence of both youth and antiquity?

One fact on which Grand Canyon geologists have always agreed is the reality of Dutton's "great denudation." Geologists from Newberry on could tell that although a sequence of Mesozoic rocks amounting to thousands of feet exists elsewhere in the Colorado Plateau, erosion has removed all those rocks from the Grand Canyon district. The volume to which a given vertical thickness of rock layers, spread over even a relatively small area, can multiply is surprising. A mile-thick sequence of rock eroded from an area 100 miles by 100 miles, considerably smaller than the Grand Canyon district, produces 10,000 *cubic miles* of sediment. The erosion, transportation, and deposition of such immense quantities of rock debris requires at least three things: streams that are energized because they flow high above their base levels; a well-integrated, long-lasting drainage system; and an equally large and long-lasting receiving basin. A young Colorado River spending most of its energy cutting down and headward rather than laterally, and bearing few significant tributaries to help carry out the erosion, seems an unlikely candidate for the main stem of such a drainage system. This line of thought caused some geologists to infer that instead, an older, more thoroughly developed, drainage system must have preceded the modern Colorado. But is there any direct, positive evidence for such an older drainage system?

The Hualapai Plateau, where Dick Young did his dissertation fieldwork, provides the critical evidence. It not only preserves a more complete Tertiary rock section than elsewhere in the Grand Canyon region, it also preserves a more ancient topography. The country of the Hualapai tells us more than any other about the conditions that confronted the early Colorado River. The most striking finding is that 50–55 million years ago, in the Eocene, the Hualapai Plateau already stood high and already contained deep canyons. Uplift associated with what geologists call the Laramide Orogeny, the mountain-building event that lifted the Rocky Mountains, produced the original elevation and relief of the Hualapai Plateau. Using the oldest rocks encountered in a well near Peach Springs Canyon as the marker of the bottom of an ancient canyon, and measuring all the way to the youngest rock exposed on its rim, Young calculated that the old channel had been at least 4,000 feet deep. Making reasonable assumptions about the rocks that erosion has likely removed from the top of the section, the depth of the canyon may well have reached a mile.

The existence of a canyon a mile deep tells us something more. As geologists starting with Newberry showed, erosion by running water has produced the topography of the Southwest, with its high plateaus and deeply etched canyons. Because a stream cannot erode below its base level, which ultimately is sea level, the presence of a stream-cut canyon 4,000 feet deep means that the region must have stood at an elevation of at least 4,000 feet. But canyons nearly a mile deep, etched into a plateau standing a mile or more high, are just what we find in the Grand Canyon today. The Grand Canyon had a grandparent, to whom it bore a distinct family resemblance.

Historically, geologists believed that erosion had worn down the Colorado Plateau region to one of William Morris Davis's peneplains. They also thought it self-evident that a plateau that now stands high and contains deeply incised canyons, but that was once a peneplain, must have been uplifted. Because the Colorado River still erodes vigorously, it seems reasonable that the assumed uplift has continued to the present, though plate tectonics offers no reason why it should. But geologists abandoned the peneplain model in general and for the Grand Canyon in particular decades ago. At the June 2000 Symposium, several vigorously disputed the evidence for modern uplift. They argued that because the Colorado Plateau already stood high during the Miocene, the subsequent collapse of the Basin and Range Province to the west could have supplied the same drop in base level as would continuing uplift of the Colorado Plateau. In their view, it is not necessary to appeal to uplift continuing to the present day to explain the incision of the Grand Canyon; the original uplift would have been sufficient.

Young also found that after the Laramide uplift, streams brought far more sediment to the western Grand Canyon district than they removed. Eventually, a blanket of debris several hundred feet thick—the rim gravels derived from the west and southwest—covered the Hualapai country. Not until the Miocene, when the Basin and Range Province crumpled and the drainage reversed direction, did the regime switch from net deposition to net erosion.

Geologists have long recognized that before downfaulting, the Basin and Range Province stood higher than the region to the east, which afterwards would become the Colorado Plateau. The regional drainage, including Young's ancient system, must then have run from west to east—from Nevada into northern Arizona and Utah and on toward Colorado—opposite to today. Powell himself recognized that the rivers must then have reversed direction:

> At last the movements which began at the commencement of Tertiary time succeeded in bringing the whole [Colorado Plateau] area not only above the level of the sea, but above the general level of the Basin province itself; so that while the Basin province was drained into the Plateau province in earlier Tertiary time, in late Tertiary time the drainage was reversed and the streams of the Plateau province found their way to the sea by passing through the Basin province.

Putting four sets of facts together—the great denudation, the ancient drainage system in the western Grand Canyon, the subsequent reversal of the streams of that system, and the lessons of plate tectonics—a new picture emerges. At the 2000 Symposium and elsewhere, several geologists presented evidence to support what I will call the Lazarus Theory.

As Powell and his successors recognized from a variety of evidence, before the central Rocky Mountains and the Basin and Range Province existed, the regional slope in the Southwest funneled waters from southwest to northeast toward some distant outlet, possibly a vast inland basin in Utah or Colorado. One paper at the 2000 Symposium described Eocene sedimentary rocks in Southern Utah that could have formed in such a basin. These streams cut deep canyons into the Plateau margin. They were the ancestors of the present Grand Canyon and may have been equally impressive, had anyone been there to see them.

The subsequent rise of the Rockies and the downfaulting of the

Basin and Range Province left the Colorado Plateau standing lower than lofty mountains to its east, but higher than the broken country to its west. This new topographic arrangement forced the rivers of the old drainage system to do an about-face and head southwest. Continuing erosion of the high country south of the Mogollon Rim blanketed the Grand Canyon region with the rim gravel debris. The water that still reached the drainages in the western Grand Canyon trickled westward through the gravel, some of the time running underground. Because there was no through-flowing, large river at this early stage, interior basin sediments could accumulate to become the Muddy Creek Formation, some time after which the Colorado River could have begun to flow through the area. Some four to five million years ago, plate movement opened the Gulf of California, providing a new outlet for the lower Colorado River. After the ancestral Colorado breached the Grand Wash Cliffs, newly energized streams began to re-excavate the ancient, gravel-filled channels and the thick sedimentary cover. Erosion eventually removed most of the gravel, integrated the streams, deepened the older canyons, and connected the western Grand Canyon with the Colorado River above the Kaibab Uplift. Though geologists have no more proven the theory of an ancestral canyon country than any of the others, it is consistent with the facts as we know them today. Regardless, we still have the question of how the modern Grand Canyon—the one we see today—came to be.

ANOTHER OLD IDEA, ALSO rejuvenated at the Symposium, is that streams on the Colorado Plateau once connected a chain of lakes. As these connecting channels eroded and deepened, they eventually drained the lakes, leaving a river flowing in a set of canyons between basins. Gilbert showed that rivers are in dynamic equilibrium, evidently able to adjust more rapidly than the timescale of most earth movements. Upset its equilibrium, and the river will take steps to restore it. Add sediment; the river will remove it. Remove sediment; the river will add it. Straighten its channel; the river will erode back toward its natural curvature. Decrease its velocity; the river will dump its sediment load. Increase its velocity; the river will erode its channel and banks to gain more load.

Dam a river, as nature and government bureaucracies have a proclivity for doing, and it will set about removing the dam. Lava flows, debris slides, glacial moraines, and the like all produce natural dams. No matter whether a dam is man-made or natural, the fate of the lake it impounds is immediately sealed. When it enters a lake, the velocity of a stream falls to zero; indeed, the stream ceases to exist. The sediment that the stream carries, which is unchanged by the existence of the dam, now settles out onto the floor of the lake, eventually filling it completely. Then the lake waters overflow through the lowest point in the dam and quickly erode and remove it. Now the river, again flowing freely, soon reestablishes its equilibrium profile, along the way removing most of the sediments deposited in the lake and perhaps leaving little evidence that a lake ever existed. Ultimately, rivers destroy lakes—that is why there are no geologically old lakes. Several eminent geologists have thought that now-vanished lakes once existed on the Colorado Plateau and might have been instrumental in creating the Grand Canyon.

Eliot Blackwelder well understood the geologic importance of lakes. As one who had come of age during the rise of the Bureau of Reclamation, the greatest dam-building bureaucracy in the history of the world, he also understood the role of megadams in the western United States. He had seen Hoover Dam rise and inundate the area west of the mouth of the Grand Canyon and create Lake Mead. In 1949, the Bureau of Reclamation issued a planning report identifying several excellent sites for the next set of big dams: Why not dam the western Grand Canyon at Bridge Canyon, Glen Canyon, and both Powell's Flaming Gorge and Echo Park, far up on the Green? The proposal was a trial balloon to see how much support and opposition would develop, but the Bureau of Reclamation got more of the latter than it bargained for. Indeed, one can trace the maturation of the

Sierra Club and the origins of the modern environmental movement back to this time. David Brower, then the new president of the Sierra Club and later John McPhee's Archdruid, made opposition to the Echo Park dam his first cause. Harvard historian Bernard De Voto wrote an angry piece for the *Saturday Evening Post* that the popular *Reader's Digest* reprinted. But it was another publication that saved Echo Park. With the encouragement of publisher Alfred Knopf, who also served as chairman of the Advisory Board on National Parks of the Interior Department, Wallace Stegner edited and wrote an essay for the 1955 publication *This is Dinosaur*, an elegant and eloquent collection of articles and photographs. The geologist chosen to contribute to the book was Eliot Blackwelder. The Bureau of Reclamation never built the Echo Park and Bridge Canyon dams (at least, they have not yet done so, environmental battles never being finally won), but in a Faustian bargain, their opponents settled for the Flaming Gorge and Glen Canyon Dams. Brower regretted the latter in particular to the end of his days, writing, "Glen Canyon died, and I was partly responsible for its needless death." By 2005 ongoing drought in the West has reduced Lake Powell to only one-third of its capacity. Glen Canyon is rapidly re-emerging from the depths, though for how long no one knows.

In his 1934 article, two decades before the Echo Park controversy, Blackwelder called attention to a way in which a set of lakes, as they inexorably engineer their own destruction, could carve a great canyon. As he acknowledged, even then the idea was not new. In his contribution to the 1861 Ives report, Newberry had described the same process:

> Doubtless in earlier times [the river] filled these basins to the brim, thus irrigating and enriching its entire course. In the lapse of ages, however, its accumulated waters, pouring over the lowest points in the base, forming that remarkable series of deep and narrow cañons through which its turbid waters now flow, with rapid and almost unobstructed current, from source to mouth.

The geologic evidence convinced Blackwelder that lake overflow and integration had helped to carve the Grand Canyon. His began his model in the late Tertiary, when the climate of the Colorado Plateau was arid. Low mountain ranges separated broad basins, as they do today in the desert Southwest, Mongolia, the Kalahari, and Central Australia. Under these conditions, drainage is interior to the basins and "even the few persistently flowing streams dwindle and end in 'sinks.'" Late in the Pliocene or early in the Pleistocene, uplift raised the Rocky Mountains, blocking the eastward-moving winds, increasing precipitation and filling the basins to the west of the mountains.

As each basin filled, it spilled over, and its escaping waters cascaded down to the next-lowest basin, eroding a channel between the two. As the process continued, the channels linking the lakes deepened until gorges separated a chain of lakes. Before long, the connecting channels had drained the lakes; eventually all that remained was a single master stream connecting a series of basins. Erosion continued until the master stream established its equilibrium profile, by which time it had removed most or all of the lake sediment. Now the Colorado "resembled somewhat its present self—a river flowing through successive broad depressions without lakes and through canyons in the ranges that separate the basins." In time, erosion erased all traces of the earlier lakes, and the Colorado River, in a "haphazard and accidental development, spilled over into the trough of the Gulf of California" and built a delta.

This is a venerable theory—Longwell cited it among his possible explanations of the Grand Canyon. But is there evidence for it, and are there modern analogues? Norman Meek, a professor at California State University at San Bernardino, east of the Los Angeles basin, believes there is at least one. Meek did his fieldwork on the Mojave River near his university, a terrain much like the one that Blackwelder envisioned and where Blackwelder himself had once worked. He and his coworkers deduced that the Mojave River originated with the uplift of the San Bernardino and San Gabriel Mountains during the

Pliocene Epoch. The river extended itself downstream by establishing lakes in the basins, then breaching the downstream rim of each basin to connect to the next, as Blackwelder had suggested for the Grand Canyon. The influx of water from melting glaciers and snowfields at the end of the last Ice Age accelerated the Mojave's downcutting. Meek was able to show that in only a few thousand years after the lakes drained, "more than 99 percent of the lake clays have eroded." Unfortunately, this means that the direct evidence for Meek's chain-of-lakes hypothesis for the Mojave has vanished, leaving him to work with circumstantial evidence, the lot of those who try to find long-gone lakes and rivers.

At the Symposium, Meek and, in a separate paper, Jon Spencer and Phillip Peathree of the Arizona Geological Survey endorsed the lake overflow theory for the origin of the Grand Canyon. Both preferred the idea to the headward erosion endorsed by most geologists today. Meek theorized that about ten million years ago, the ancestral Colorado River rose in the Rocky Mountains and flowed downstream through a series of broad basins. The critical one of these basins contained Hopi Lake, its former presence now revealed by the sedimentary rocks of the Bidahochi Formation (see Figure 9, page 193). Meek said that Hopi Lake might have filled to overflowing and spilled west, heading for its base level at the bottom of the Grand Wash Cliffs and starting the incision of the Grand Canyon. By four to five million years ago, the river had reached the downstream basins where the Bouse and Imperial Formations accumulated. Integrating these basins downstream one at a time by spillover and erosion, the river eventually found its way on to the Salton Trough and the Gulf of California. After the sediments were removed, all that was left was a river flowing above the Grand Wash Cliffs in a deep canyon and below the cliffs through a set of structural basins.

At the time of the McKee symposium in 1964, geologists could only say that the Bidahochi Formation, deposited in the putative Hopi Lake, is "Miocene-Pliocene." But at the June 2000 Symposium, a group presented new results based on argon-argon dating of volcanic ash beds throughout the Bidahochi sediments (see Figure 9, page 193). The dates extended from sixteen down to seven million years. This meant that Hopi Lake would have been around at the right time for the lake integration theory, just before the modern Grand Canyon began to appear. But the theory also requires Hopi Lake to have had a large enough volume so that when it drained, the water released could have accomplished a great deal of erosion. Unfortunately, the present evidence suggests that Hopi Lake, like most desert lakes, was shallow, even ephemeral. Meek speculates that a deeper lake later formed atop the Bidahochi sediments, which erosion later removed. Thus, lake overflow and integration appears to be another speculative idea—an educated geologic guess—without direct evidence. Yet one ought not be too quick to reject an idea suggested by geologists of the stature of Newberry, Blackwelder, and Longwell. Perhaps lake spillover and integration involved not Hopi Lake, but other lakes, the evidence for whose existence erosion has removed.

As is abundantly clear by now, unraveling the history of something as complicated as the Grand Canyon is not a matter of proving one theory right and all others wrong. The development of the Grand Canyon is so complex, and the evidence so scanty, that proof is hard to come by. Unlike most other scientists, geologists cannot conduct experiments to solve their problems. Instead, they must work with what Nature has happened to leave for them to find, if anything. Yet whatever frustration derives from having to decide to what extent the absence of evidence constitutes evidence of absence, most geologists would judge that being able to spend a career in Grand Canyon country more than makes up for it.

The influential philosopher of science, Karl Popper, argued that scientists never actually prove a theory. At any moment, scientists may discover some important new fact that a prevailing theory cannot explain and which therefore falsifies the theory. Popper is correct that scientists never finally prove a theory; after all, Einstein's theories superseded even those of the great Newton, and one day others may supersede Einstein's. As to the notion that a single, contradictory fact falsifies a leading theory, nothing could be further from the truth. To send a theory to the dustbin requires not just one fact, but many; they take time to accumulate. It also usually takes more than one generation for human nature to adjust. German physicist Max Planck was not far wrong when he said:

> An important scientific innovation rarely makes its way by gradually winning over and converting its opponents. . . . What does happen is that its opponents gradually die out and that the growing generation is familiar with the idea from the beginning.

Grand Canyon geologists have not proceeded by disproving theories. Nor, as we have seen, have they proved them. Rather, over the 140 years from Newberry to the June 2000 Symposium, they have drawn the boundaries ever tighter around the most likely theories, aided recently by the results of radiometric dating. Gradually, some older theories, like those of Powell, Dutton, Hunt, and McKee, have come to appear sufficiently unlikely that they can be set outside the boundaries of the probable. Geologists then pursue the more likely theories, which turn out to be either flawed or worthy of follow-up.

Being human, scientists need encouragement. An absence of evidence provides none. Partly for this reason, theories for which no new, positive evidence is forthcoming gradually recede in interest. But, unsatisfying as it may be, scientists have not falsified the older theories in Popper's sense. Rather, scientists have put them on the shelf, where a single, solid piece of new supporting evidence—a key river gravel deposit in just the right place—could rejuvenate them like a stream on an uplifted plateau.

There remains one critical kind of rock to discuss, the youngest found in the Grand Canyon. We find evidence of them in the fantastic basalt lava flows that spill over the rim of the western Grand Canyon like frozen rock waterfalls. They tell of a time of intense volcanic activity almost as hard to comprehend as the Big Cañon was for Cardenas and his men. With these lavas, geologists do not have to worry about evidence of absence conundrums—they hold the specimens in their hands and can date them absolutely. These new results are causing a revolution in our understanding of how fast the river incised its Grand Canyon.

Chapter 15

Molten Rock,
Melted Snow

One can only marvel at how John Wesley Powell could greet with apparent equanimity sights that neither he nor any other geologist had seen or even imagined. One of the most marvelous confronted him in the Toroweap section, some ninety miles downriver from the present Grand Canyon National Park Headquarters. By this point in their journey, late August of 1869, he and his men were down to the barest of rations. The main thing on their minds was which would end first: their food or the Grand Canyon. But showing his usual mixture of sobriety and optimism, Powell wrote, "It is curious how anxious we are to make up our reckoning every time we stop, now that our diet is confined to plenty of coffee, very little spoiled flour, and very few dried apples. It has come to be a race for a dinner. Still, we make such fine progress, all hands are in good cheer, but not a moment of daylight is lost." In the Toroweap stretch and downstream, the volcanic geology was so spectacular that for a while, at least, it seemed to have made Powell forget his growling stomach. Judging from the journal entry of crew member Bradley, this was all too typical: "If he can only study geology he will be happy without food or shelter but the rest of us are not so afflicted with it to an alarming extent."

Here is how Powell described his wonderment:

Great quantities of cooled lava and many cinder cones are seen on either side; and then we come to an abrupt cataract. Just over the fall; on the right wall, a cinder cone, or extinct volcano, with a well defined crater, stands on the very brink of the canyon. This, doubtless, is the one we saw two or three days ago. From this volcano vast floods of lava have been poured down into the river, and a stream of the molten rock has run up the canyon, three or four miles, and down, we know not how far. Just where it poured over the canyon wall is the fall. The whole north side, as far as we can see, is lined with the black basalt, and high up on the opposite wall are patches of the same material, resting on the benches, and filling old alcoves and caves, giving to the wall a spotted appearance. The rocks are broken in two, along a line which here crosses the river, and the beds, which we have seen coming down the canyon for the last thirty miles, have dropped 800 feet, on the lower side of the line, forming what geologists call a fault. The volcanic cone stands directly over the fissure thus formed. On the side of the river opposite, mammoth springs burst out of this crevice, one or two hundred feet above the river, pouring in a stream quite equal in volume to the Colorado Chiquito. This stream seems to be loaded with carbonate of lime, and the water, evaporating, leaves an incrustation on the rocks; and this process has been continued for a long time, for extensive deposits are noticed, in which are basins, with bubbling springs. The water is salty. We have to make a portage here, which is completed in about three hours, and on we go.

We have no difficulty as we float along, and I am able to observe the wonderful phenomena connected with this flood of lava. The canyon was doubtless filled to a height of twelve or fifteen hundred feet, perhaps by more than one flood. This would dam the water back; and in cutting through this great lava bed, a new channel has been formed, sometimes on one side, sometimes on the other. The cooled lava, being of firmer texture than the rocks of which the walls are composed, remains in some places; in others a narrow channel has been cut, leaving a line of basalt on either side. It is possible that the lava cooled faster on the sides against the walls, and that the centre ran out; but of this we can only conjecture. There are other places, where almost the whole of the lava is gone, patches of it only being seen where it has caught on the walls. As we float down, we can see that it ran out into side canyons. In some places this basalt has a fine, columnar structure, often in concentric prisms, and masses of these concentric columns have coalesced. In some places, when the flow occurred, the canyon was probably at about the same depth as it is now, for we can see where the basalt has rolled out on the sands, and, what seems curious to me, the sands are not melted or metamorphosed to any appreciable extent. In places the bed of the river is of sandstone or limestone; in other places of lava, showing that it has all been cut out again where the sandstones and limestones appear; but there is a little yet left where the bed is of lava. What a conflict of water and fire there must have been here! Just imagine a river of molten rock, running down into a river of melted snow. What a seething and boiling of the waters; what clouds of steam rolled into the heavens!

Thirty five miles today. Hurrah!

Until near the end of the twentieth century, the only way to see the lavas of the Toroweap was like Powell, by floating past them, or like Dutton, by gazing down on them from the rim. The first method allows too little time on site; the second leaves the lavas too far away. Today, even to get to the Toroweap Overlook, one has to drive sixty miles of gravel road from Kanab, Utah or ninety miles from St. George, where one then perches thousands of feet above the lavas deep in the canyon. The view is worth the trip, but much of the geology remains out of reach.

Geologist William K. Hamblin of Brigham Young University decided to solve the problem by setting up a series of base camps on the canyon floor and supplying them by boat from Lees Ferry. That way, he and his crew could remain in the canyon for weeks at a time, working their way downstream as the field season progressed. Even so, many of the basalt flows hang so precariously that Hamblin could only photograph them from a distance or peer at them from a helicopter.

Hamblin found that more than 150 separate lava flows have poured over the rim and down into the canyon:

Some flows were extruded on the Uinkaret Plateau and cascaded over the north rim of the canyon into Toroweap Valley and Whitmore Wash. Others were extruded within the canyon itself and spread out over the Esplanade Platform before forming spectacular frozen lava falls that plunged over the rim of the inner gorge into the Colorado River 3000 feet below. In several places, volcanoes are perched precariously on the very rim of the canyon, and remnants of others cling to the steep walls of the inner gorge.

Because all that survives of most of the flows is a thin, vertical slab of basalt plastered to the canyon walls, Hamblin's detective work is all the more remarkable. Sometimes the slabs are stacked side-by-side, each younger than the one to the inside, standing Steno's Law of Superposition on its side.

As Hamblin began to work out the history of the lava flows, what emerged seemed closer to science fiction than reality. From direct observation in Iceland and Hawaii, we know that basalt can flow forty miles an hour and travel long distances. Hamblin found that one of the Toroweap basalt flows had run down the floor of the Grand Canyon for eighty-five miles! The flows were thickest where they landed on the canyon floor. Downstream from there they tapered off, forming a giant wedge that dammed the river completely, with the thick end upstream. The dams were hundreds of feet high. As the Pleistocene glaciers and ice fields melted far away in the Rocky Mountains, the flows reaching the dams may have ballooned to all-time-high discharges. One geologist estimates that the volume of water reaching the Grand Canyon from upriver may have exceeded one million cubic feet per second (cfs), compared with the highest ever measured on the modern Colorado, 300,000 cfs in 1884.

Any river rafter, geologist or not, can see the lava flows in the Toroweap and plasters of basalt and protrusions of basalt here and there downstream for scores of miles. There is no doubt that the lava dams existed—but for how long? Hamblin concluded that they had lasted long enough to impound giant lakes that cascaded over the dams in correspondingly large waterfalls. According to his calculations, one dam impounded a lake larger than Mead and Powell combined; it extended from Toroweap more than 200 miles upstream, all the way to Moab, Utah. Descending their Grand Staircase, Dutton's Titans would have found a bathtub big enough to fit them. This putative reservoir would have taken twenty-two years to fill with water and reached up to the level of the Redwall Limestone (see Figure 16, page 260), halfway up the Grand Canyon—and 3,000 years to fill with the sediment brought to it by the muddy Colorado. (As we will see in a moment, these lakes are millions of years younger, and in a different place, than Hopi Lake and Peach Springs Lake.) Hamblin envisioned one waterfall as over 2,000 feet high, fifteen times higher than Niagara, and says it would have discharged at 4.8 million cfs, likely making it the largest waterfall in earth history. But unfortunately for this exciting image, more recent work places the height of the falls at only one-tenth of Hamblin's estimate.

Water topping the upper end of one of the wedge-shaped dams would have rushed down to the lower end, where a near-vertical face of basalt lay atop loose river sand and gravel. The sudden immersion of the hot lava in cold water would have left it brittle and partly shattered. Moreover, the lava contained vertical cracks into which water could seep and press outward. These conditions caused the dam to erode and quickly retreat upstream. As it did so, the wedge shape of the dam caused the height of the waterfall to rise, giving the rushing waters more energy. That in turn caused the falls to retreat faster, and so on, until, in one giant crescendo of rock and water, the upstream end of the wedge suddenly collapsed. Now the river flowed across a sea of mud and loose sediment, which it quickly removed. Soon, all that remained of the dam and its lake was the odd slab of basalt stuck to the canyon wall.

Massive, near-instantaneous dam failures are not fictional. Unfortunately, America has two prime examples: the Johnstown Flood in Pennsylvania and the Teton Flood in Idaho. In 1889, a poorly constructed and maintained earthfill dam above Johnstown broke apart during heavy rains, sending a wall of water downstream on the town at speeds up to forty miles per hour. More than 2,200 people died in an event that helped give rise to the Bureau of Reclamation. The second collapse took place in Idaho in June 1976. Before engineers had completed its outlet works, unusually high runoff from melting snow in the Teton Range began to fill the new, earthen Teton Dam. With almost no warning, the dam failed catastrophically in a few hours, sending 80 billion gallons down in a wall of water 20-30 feet high, onto the unsuspecting Mormon towns below, some of which disappeared from the map of Idaho. The collapse of the largest lava dams in the Grand Canyon would have had a much greater effect. Geologist Robert Webb of the University of Arizona and Cassandra Fenton of the University of Utah estimate that the failure of the lava dams produced flows as high as 15 million cubic feet per second—thirty-seven times the largest known Mississippi River flood.

Geologists do find the expected rubble from the collapsed lava dams, but not the corresponding sediment that should have accumulated in the giant lakes, small patches of which should still be present here and there. What had appeared to be lake sediment turns out to be something else. Thus, at present, there is evidence of the existence and destruction of the dams, but not that they lasted long enough to fulfill Hamblin's vivid conception. Nevertheless, his work remains of major importance, for it reveals much about how rivers evolve: just as Gilbert thought.

Using expert field geology and potassium-argon dating, Hamblin worked out the age, thickness, and length of each dam. He found that if the dams had been stacked atop each other, their total thickness would have exceeded 11,000 feet. Of course, the lavas were not stacked atop each other; rather, they appeared episodically: as soon as one solidified, the river removed it; then some time passed before the next flow erupted and the process repeated, and so on. Hamblin estimates that basalt was actually present and undergoing erosion for a total of about 200,000 out of the last million years. To slice through more than two miles of hard basalt in one-fifth of a million years means that the Colorado has "the capacity to erode through any rock type almost instantaneously." Regardless of the exact accuracy of Hamblin's estimates, one can see that the Colorado River has entirely removed the dams that once existed, leaving only scattered slabs of basalt here and there. Thus, in only a moment of geologic time, melted snow trumps once-molten rock.

Obviously, each lava dam temporarily upset the river's approach to its equilibrium profile. The river immediately began to retaliate by eroding its way through the barrier. After the river returned to the profile it had before the dam interrupted, it essentially stopped eroding downward. This is testament to the great erosive power of the Colorado River and to Gilbert's concept of a graded stream. Not only did the river cut through a mile of solid rock, not only did it remove every obstacle in its bed, but before the dams interrupted it, the Colorado River had reached its equilibrium profile. And after removing the lava dams, it regained that profile almost immediately.

A river's actual downcutting can take place only where water is in contact with rock—in the river's bed and along its immediate banks. Without some other process at work, the Grand Canyon would be an abyssal slot canyon thousands of feet deep but only 200-300 feet wide— the width of the river itself. Thus, the erosion that has made the Grand Canyon such a spectacular sight is due not to the direct action of the Colorado River, but to the weathering and downslope movement of material that has caused the slopes on either side of the canyon to retreat. How rapidly do such side slopes move back? Hamblin found that no sooner had erosion removed a dam than nearly all the lava that had plastered itself onto the canyon walls also quickly disappeared and the slope returned to the shape and position it had before the lava dam blocked it. But, just as the river returned to its equilibrium profile but no farther, so "the process of slope retreat did not enlarge the canyon and go back beyond the original canyon walls." Gilbert would have been grateful to Hamblin, but not in the least surprised.

TO DATE THE BASALTS OF THE western Grand Canyon, Hamblin collaborated with Brent Dalrymple, a pioneer in the use of potassium-argon dating and one of the inventors of the paleomagnetic timescale. Geologists began to develop the potassium-argon and the other radiometric methods during the 1950s. Each depends on knowing three things: the amount of a parent isotope in a rock or mineral specimen, the amount of the daughter isotope, and the rate at which the parent decays into the daughter. Knowing those three, scientists calculate the time that the parent has spent changing into the daughter. The reliability of the methods depends on three assumptions: first, that none of the daughter isotope was present when the rock or mineral formed (no original daughter). Second, that the specimen has neither lost nor gained parent or daughter (closed system). Third, that the rate of radioactive decay has been constant.

Because no scientist has ever detected a significant variation in the rate of radioactive decay, no matter the heat and pressure to which they subjected the parent isotope, we can consider the third assumption validated. But the first two assumptions represent more serious obstacles to accuracy, for they do not always hold. At first, the failure of these assumptions appeared to be fatal to radiometric age dating. But over time, geologists have invented clever ways either to avoid the assumptions or to recognize when a given set of specimens fails to meet them. In the latter case, the geologist discards the results and looks for new specimens that do obey the assumptions. Sometimes none do, and the scientist has to give up on dating that particular rock.

Geologists have long known that with the potassium-argon method, the supposition that no original daughter isotope was present sometimes fails because some minerals, when they crystallize, trap the argon that is present in the earth's atmosphere. These minerals wind up having more daughter isotope than they should and therefore appear older than they are. Volcanic glass like obsidian is especially apt to trap gases, and though geologists try to avoid specimens that contain glass, its presence on a microscopic scale can be hard to rule out. Oppositely, minerals may lose argon when they subsequently heat up, causing them to have less daughter isotope than they should and to appear younger than they are. But geologists generally can tell from other evidence when metamorphism has reheated a rock. In any case, young volcanic rocks resting on the surface, like Hamblin's basalts, have had no opportunity to re-heat. That they may have trapped inherited argon is a more serious worry.

The strongest evidence of the reliability of the potassium-argon method generally is that geologists used it exclusively to establish the paleomagnetic timescale, on which the theory of plate tectonics rests. On the other hand, hundreds of potassium-argon measurements went into constructing that timescale, providing strength in numbers. A single potassium-argon date on a single rock or mineral is not as reliable as geologists would wish.

To date young volcanic rocks with greater accuracy, geologists needed a radiometric technique that avoids or minimizes the problem of inherited argon. They came up with a method they call argon-argon, because even though it depends on the decay of potassium to argon, surprisingly, there is no need to measure the amount of parent potassium. In the first step, geologists expose the potassium in the specimen to radiation in a nuclear reactor and convert it into a unique isotope of argon. Next, the geologist heats up the specimen enough to drive off the trapped argon, which binds less tightly in crystal structures. This leaves only the radiogenic argon and that converted from potassium. Scientists can easily tell the different isotopes of argon apart in a mass spectrometer. They use the amounts of these isotopes to work back to the amounts of potassium and argon in the original specimen and calculate its age. The method is self-correcting in that geologists can tell from their raw data when they failed to drive off all the trapped argon.

How well does the argon-argon method work? That question is really two: how young a rock can the method date, and how precise (replicable) is the method? A group of researchers addressed the first question by attempting to date an exceedingly young volcanic rock of well-known age: the 79 A.D. eruption of Vesuvius, which destroyed Pompeii. At the time of their work, the eruption had taken place exactly 1,918 years before. In a paper titled "Calibration against Pliny the Younger," the team reported that the inherited argon in the samples would have caused the conventional potassium-argon method to give too old an age. Instead, using the argon-argon method and first driving off the inherited argon, they got 1,925 ± 94 years.

Another test of the method's precision comes from the argon-argon dating of the Bidahochi Formation, whose sediments include layers of volcanic ash interspersed at frequent intervals. According to Steno, the lowest ash bed is the oldest and the highest is the youngest; those in between should have intermediate ages, with younger always above older. The following table shows the measured ages of the ash beds listed in order of position in the section:

Position	Age in Millions of Years
Highest	6.6
	7.7
	13.71
	13.77
	13.78
	15.19
	15.46
Lowest	15.84

Geologists test replicability by using several measurements on the same specimen, or on a set of closely related specimens, and seeing how well the dates match up. One of the most striking of the lava dams in the Grand Canyon is the Black Ledge flow exposed between river miles 207 and 209 below Lee's Ferry. Several authors have dated a total of seven specimens from the flow, with results ranging from 520,000 to 620,000 years. The most recent measurements have error bands of only 1-2 percent. Because the scientists did not date the same part of the flow, the differences are likely real. All these dates are in the same range and therefore speak to the replicability of the method.

The Grand Canyon basalts reveal that the area was at least as volcanically active as today's Iceland or Hawaiian Islands. No one was around to see it, but to witness floods of red-hot lava spilling over the lip of the Grand Canyon and plunging down to die, hissing and steaming, in the cold waters of the Colorado River, would have been one of the most spectacular of sights in earth history. At first, the molten rock would have appeared to still and vanquish even the mighty Colorado.

But the conquest of water by rock was only apparent. All too soon, the dams collapsed and disappeared, and running water had again demonstrated the superiority that geologic time provides it.

EVEN IF THE POSSIBILITY explored in the last chapter is correct—that an ancient drainage system reversed itself to exhume an ancestor of the modern Grand Canyon—we still need to ask: after the exhumation was over, how rapidly did the Colorado River incise the canyon that we see today? Geologists measure the rate of incision using a lava or other deposit that was once at river level, but which continued erosion and lowering of the bed of the Colorado have today left standing well above the river. The present height of the lava flow above the river, divided by the flow's measured age, gives the rate of downcutting. But how can we know that lava actually flowed out at river level? One telltale sign is that when lava comes in contact with water, it freezes into rounded shapes called pillows. Some of the Grand Canyon flows have pillows on their lower sides, showing that they spilled into the ancient river. Another sign is to find lava intermingled with or resting on river gravels. Geologists have found both kinds of evidence in the Grand Canyon and dated several of the once-riverside basalts.

Most of this work has focused on measuring the incision rate above and below the giant Toroweap and Hurricane Faults (see Powell's cross-section diagram, Figure 3, page 48), which cross the river at mile 179 and mile 188, respectively. The measured incision rate downstream of the faults is roughly 70 meters (230 feet) of downcutting per million years—but curiously, the rate upstream is twice that. The Grand Canyon never seems to run out of puzzles: why would the Colorado River have cut the eastern section, upstream of the big faults, twice as fast as the western section?

The answer is that the downstream, western side of both faults dropped relative to the upstream side, leaving the downstream side standing closer to its base level and the upstream side higher above it. In order to maintain the uniform grade that it clearly displays, the Colorado River has had to cut through more rock in the relatively higher upstream section, and cut it faster, than in the relatively lower downstream section.

But a back-of-the-envelope calculation gives another curious result. The measured incision rate upstream from the faults is about 140 meters (460 feet) per million years. In the five to six million years that geologists believe the Colorado River has been eroding the Grand Canyon, that rate would produce a gorge at most 840 meters (half a mile) deep. Yet the eastern Grand Canyon is often twice that deep. Some other process must have enhanced the downcutting.

Ivo Lucchitta offers three possible explanations for the discrepancy between calculated incision rate and the actual depth of the canyon; they might have operated together. First, a knickpoint migrating up the Lower Colorado River, stimulated by the opening of the Gulf of California, may have reached and crossed the Grand Wash Cliffs about six million years ago, then continued its way upriver well into Utah. The erosion produced by this knickpoint would have added to that induced by the change in base level generated by movement on the Hurricane and Toroweap faults. Second, as Lucchitta has consistently argued, continued uplift of the entire Colorado Plateau may have kept dropping the base level of the plateau streams and reinvigorating them. As noted earlier, the Colorado Plateau already stood high long before the Grand Canyon appeared. Some geologists believe there is little independent evidence of continued uplift. Third, capture of the ancestral upper Colorado by the lower, as McKee and company outlined, greatly increased the volume of water flowing in the river west of the Kaibab and thus enhanced its downcutting power. The greatly increased flow from Pleistocene rainfall and melting snow and ice would have added even more volume.

These are familiar processes by now, but Joel Pederson of Utah State University and his colleagues came up with two other possible explanations of why the present rate of incision could not have carved the Grand Canyon in five to six million years. One is the simple idea that, logically, the rate of incision must decline over time, in which case one cannot use the present rate in a historical calculation. After a sudden and drastic fall in base level, a rejuvenated river will begin to cut down rapidly. But the more the river cuts down, the nearer it gets to its base level, assuming there is no further uplift. The closer a river is to its base level, the less rapidly it erodes. This is the essence of Davis's model of aging landscapes (see Figure 6, page 155.) Thus, the river must have incised more rapidly in the past than recently, and the present rate must be less than the long-term average. Pederson's second explanation is music to the ears of those who believe that an old canyon has been exhumed: "some canyon cutting . . . happened prior to the Colorado River taking its full present path at six million years and creating the Grand Canyon we see today." If the modern river does not cut fast enough to have carved the modern canyon, perhaps an ancestral river provided a head start.

Whatever the exact explanation, an important point remains: although the upstream sides of the two large faults have risen relative to the downstream by nearly 2,000 feet, nowhere in the river bed is there a bedrock rapids, much less a waterfall. Where the river runs across the trace of the Toroweap and Hurricane faults, it has removed all sign of them, showing that the river erodes more rapidly than the great faults move. In spite of every obstacle that Nature can place in its path, the Colorado River evidently has the power to achieve and maintain its equilibrium profile.

The giants of Grand Canyon geology are long gone now. Powell, Dutton, and Davis may have been wrong as often as they were right about the history of the Colorado River and its canyon. Yet without their shoulders to stand on, the geologists who came after would not have been able to see as far and would have been led down even more false trails. The progeny of the pioneers are hard at work today on the venerable puzzle, using tools that their forefathers could never have imagined. The training, methods, and equipment of these modern plateau scientists steadily improve, producing a continual accumulation of new evidence and greater expertise. Clearly geologists grow ever closer to finding the solution to their grandest puzzle. Knowing all this, an author is tempted to wait for more evidence, and then more, and so on, before summing up. But to procrastinate at this stage would be unfair not only to the dedicated reader, but also to the geologists who have done the work of the latter part of the twentieth century. Let us instead see how far we can go toward answering the question with which we began: what caused the Grand Canyon?

Chapter 16

What Caused
the Grand Canyon?

~~~~~~

The question of the origin of the Grand Canyon really has two parts: First, what were the ultimate causes of the Grand Canyon? What large-scale forces brought it about? Second, exactly how did the Grand Canyon form? Given those ultimate causes, what sequence of events—what processes and in what order—led to it? Paraphrasing the famous query in the Watergate investigation, this second question comes down to the following: "Where did the Colorado River flow, and when did it flow there?" But because deciphering the history of a river that shreds its records is even more difficult than exposing Washington cover-ups, sometimes we have to look for sediments like those of the Muddy Creek Formation and use them to answer the question, "Where did the river not flow, and when did it not flow there?"

In attempting to sum up, the first thing to note is that no matter how frustratingly difficult, the effort to understand the Colorado River and its Grand Canyon has been well worth it. As I will explore later on, the more counterintuitive is the solution to a problem, the more we learn from it. It is the surprising, the unexpected, the serendipitous discovery that leads to the greatest progress. One can draw an analogy with medical researchers, who often find that they learn more from extreme examples than normal ones. By coming to better comprehend the Grand Canyon—an extreme example of stream erosion and canyon cutting, if there ever was one—geologists have learned lessons that apply to all valleys and the streams that carve them.

In addressing the cause of the Grand Canyon, ideally we would start at the beginning and trace the evolution of the Colorado River and plateau topography forward in time. But because of erosion and other geologic processes, the farther back we look, the less evidence and the more gaps in the records we find. In the Grand Canyon region, we can go back no further than Young's Hualapai Plateau paleocanyons. The landscapes that existed before that time are lost to us.

We do know that the fossil canyons are the product of the Laramide mountain-building event that began 65 million years ago and may have persisted to the end of the Oligocene. The Laramide Orogeny not only uplifted the Rocky Mountains, it also created the Mogollon Highlands of Central Arizona and raised the Colorado Plateau at least a mile above sea level. The elevation of the Laramide-age terrain endowed its streams with the energy to cut deeply incised canyons that drained into basins to the north. Peach Springs Canyon and others like it on the Hualapai Plateau adumbrate this ghostly landscape, the progenitor of today's canyon country.

Thus, in seeking ultimate causes, we might first ask "What caused the Laramide Orogeny?" Until the 1960s, no one knew; now we understand that the cause was the subduction of the Farallon Plate under the western edge of the American Plate (see Figure 14, page 215). After that, continued erosion off the Mogollon Highlands eventually produced so much erosional debris that not only did it fill these ancient drainage channels, by the mid-Eocene, several hundred feet of rock rubble—the rim gravels—buried the entire surface of northern Arizona. This period was marked not by a great denudation that would signify the presence of a large, integrated stream system, but by a great deposition. No big river could have flowed through the western Colorado Plateau until erosion had largely removed the rim gravels. The Grand Canyon lay tens of millions of years in the future.

The next major event, also driven by plate motion, was the down-faulting and collapse of the Basin and Range Province. The process peaked in the mid-Miocene and left the Colorado Plateau standing thousands of feet higher than its western neighbor, yet thousands of feet lower than the Rockies to the east. Long ago, Major Powell recognized that this topographic arrangement meant that, "while the Basin province was drained into the Plateau province in earlier Tertiary time, in late Tertiary time the drainage was reversed and the streams of the Plateau province found their way to the sea by passing through the Basin province." From a thousand miles away, melting snow supplied the water required for deep incision; next door, the collapsed Basin and Range country supplied the equally important drop in base level.

In his introduction to the June 2000 Symposium monograph, editor Dick Young highlighted a puzzle that arises at this point in the chronology. As explained in the last chapter, there is strong evidence that the Colorado River cut the Grand Canyon in only the last few million years. The more geologists have studied rivers around the world, the more they have been impressed by just how rapidly a fast-flowing stream can cut down. By the mid-Miocene, the streams on the Colorado Plateau were high above their base level and seemingly poised to erode rapidly, carve canyons and dissect the terrain. Yet as far as we can tell, they did not. Something prevented the rapid incision that geologists have learned to expect.

There is a related puzzle. Between the mid-Miocene, say fifteen million years ago, and until after the deposition of the Hualapai Limestone, say six million years ago, the ancestral Colorado River seems to have disappeared. None of the plateau specialists has been able to find it. Hunt took it southwest through Peach Springs Canyon, but even at the time he wrote, Dick Young had already shown that the canyon ran in the opposite direction. The McKee report took the river southeast down the valley of the Little Colorado River and on to meet the upper Rio Grande. But no evidence has ever turned up to support the idea. Ivo Lucchitta proposed that after the river had crossed the Kaibab, it ran northwest into Utah and Nevada. But again, geologists have not found the sediments that would document the route Lucchitta supposes the ancient river to have taken. At the risk of pushing the Watergate analogy too far, we find that the geological tape recording of Colorado River history has a gap of nine or ten million years. We can erase magnetic tape, but large rivers cannot simply vanish—can they? I will come back to this mystery later; for now, let us return to more solid ground.

The interior basin deposits of the Muddy Creek Formation at the mouth of the Grand Canyon, capped by the Hualapai Limestone, show that as recently as six million years ago, the ancestral Colorado River had not yet breached the Grand Wash Cliffs and connected the upper and lower rivers. Indeed, the lower river did not exist, for plate movement had not yet opened the Gulf of California. But by five million years ago, the river flowed across the plateau, down and out through the Grand Wash Cliffs, and on to a series of downstream basins or lakes, in which the Bouse and similar formations had begun to accumulate, and from there on to the Gulf of California. These basins gradually linked up, as Newberry and Blackwelder envisioned. Before long a continuous river channel stretched north from the Gulf to the Grand Wash Cliffs.

The stage was now set for extreme canyon cutting. The western Grand Canyon region, an upland a mile or more high, stood only a few miles laterally from the much lower Lake Mead area. This juxtaposition gave the rivers of the western Grand Canyon a gradient of at least fifty feet per mile—100 times the average drop of the Nile or the Mississippi. Even today, with the Colorado at its equilibrium profile in the Grand Canyon, its gradient sometimes exceeds fifteen feet per mile, obliterating incipient waterfalls and giving river runners the white water they seek.

According to Lucchitta, the opening of the Gulf of California created a knickpoint that moved steadily upstream to the Grand Wash Cliffs, breached them, and after five million years reached Lee's Ferry at the head of Marble Canyon. The addition of the voluminous Pleistocene waters gave the ancestral Colorado River even greater downcutting ability. Everything conspired to cut a deep canyon.

But then Nature, as if determined to undo what she had done, within the last half-million years poured molten lava over the rim, filling the canyon and blocking the river. The effort failed, for the river quickly removed the lava dams and regained its equilibrium profile.

But not every geologist would agree with this general scenario. To put it more accurately, it is doubtful that any two Grand Canyon geologists would agree on the details of any scenario of how the canyon came to be. Never mind; geologists have discovered the most important fact about the Grand Canyon. Contrary to the logical deduction of the pioneer geologists, the widest, deepest canyon they had seen did not take a proportionately long time to form. The Colorado River did not begin to cut canyons in the aftermath of the great Eocene lakes and then persist to the present, as Powell and Dutton supposed. Although the Lazarus Theory might be correct and today's Colorado River might have had a distant ancestor, the modern Grand Canyon that we see today is young—no more than a few million years old.

But we still are left with the mystery of where the ancestral Colorado River ran between about fifteen million years ago and about five or six million—between the time the Basin and Range fell and the Gulf of California opened. The river apparently did not flow southwest, southeast, or northwest. In this time interval, after the drainage reversal caused by the collapse of the Basin and Range, it did not flow northeast. When faced with such a conundrum, it is a good idea to re-examine our assumptions.

One assumption is that the ancestral Colorado River west of the Kaibab was a large, perennial, through-flowing, sediment-laden river like the one today. Because such a river would have been bound to leave gravel deposits, this assumption required geologists to try to find those deposits and thus determine where the ancestral Colorado River ran, but in vain. Could the failure to find the river deposits mean that the assumption is wrong? Joel Pederson, hearkening back to Hunt, speculates that the old Colorado River did arrive west of the Kaibab, but died out by infiltrating into the porous Paleozoic limestones that are such an important part of the Grand Canyon geologic section. Far downstream, the underground waters emerged from springs. Of course, this idea requires first that a major desert river can dry up and sink out of sight and second that the Grand Canyon region show evidence of old spring deposits. As to the first requirement, several large modern rivers begin in wet highlands but end before they reach the sea in rubble, sand, and salty puddles. These include the Okavango River in Southwest Africa; the Logone in Chad, which feeds dry Lake Chad; the Tarim, which ends in the Gobi; and other rivers in east-central Australia that infiltrate into the Simpson Desert. Large rivers can simply disappear. As to the second requirement, the Grand Canyon today does have numerous examples of groundwater springs. The lower Grand Canyon and the area around Lake Mead have extensive travertine deposits, formed when water rich in calcium carbonate evaporates. Where do such waters get their calcium carbonate in the first place? By dissolving limestone.

The other, more fundamental assumption has pervaded the history of Colorado Plateau geologizing. Led by Charles Hunt, most Grand Canyon scientists have accepted that well before fifteen million years ago—before the mid-Miocene—the ancestral Colorado River above the Kaibab, in Utah and Colorado, already existed. Blackwelder was an exception, writing, admittedly a long time ago, "All geologists appear to have assumed that the river has existed continuously since Miocene, if not Eocene, time. This was, indeed, a natural assumption to make, but apparently there is almost no evidence to support it. There is, in fact, much definite testimony against it." Because geologists have not found where the river ran west of the Kaibab up to the time it breached the Grand Wash Cliffs, five or six million years ago, Hunt's multimillion-year gap in river history arises. But just how strong is the evidence that the upper Colorado is old? After all, we have spent much of this book illustrating how difficult it is to pin down the age of a river. Because the McKee Symposium focused on the Grand Canyon, their report provided no direct evidence about the age of the upper river, saying only:

> At least two different drainage systems, separated by the Kaibab upwarp, are believed to have been present. The eastern system, here called the upper Colorado drainage system, is inferred to have flowed southward from Utah and then southeastward across northeastern Arizona along the present Little Colorado River course, but in a reverse direction. The presence of this major drainage system is required during this stage to account for the vast amounts of Cenozoic and Mesozoic strata removed from southeastern Utah along the course of the present Upper Colorado River.

Some drainage system must have removed those vast amounts of strata, but did the main stem of that system have to be the ancestor of today's Colorado River? Must it have flowed south and southeast? Admittedly, these are reasonable inferences, but are they anything close to certainties? Perhaps as Blackwelder suggested, after the great denudation a set of poorly integrated streams ending in sinks, like those mentioned previously, marked the region above the Kaibab. These basins might subsequently have filled and spilled over, causing the rivers to work their way progressively downstream, as Blackwelder and Longwell had proposed.

Hunt, of course, did focus on the entire Colorado Plateau, starting high in the Rocky Mountains at the headwaters of its rivers. His self-described hypothesis contains many reasonable interpretations and inferences, but his writings also reveal how hard it was to be certain about the age of sedimentary rocks when neither fossils nor radioactivity could date them.

Hunt did interpret the evidence to indicate that some of the deep canyons of the Colorado River in its eponymous state are old. If it is not unfair to take a single example, Hunt describes the canyon of the Colorado River near Glenwood Springs as "probably cut during the Pliocene," which at the time he wrote, geologists thought began 13 million years ago and lasted until 1.8 million years ago. But a group of authors at the June 2000 Symposium, using modern argon-argon dating, write that "more than half of the entire 1270 m [4,130 feet] deep [Glenwood Canyon] was cut in the last 3 million years." In the several million years before that, they concluded, the river had cut hardly at all—might it not yet have existed? Hunt himself admitted, "The difficulties of interpreting the geologic history of a major river system, the vastness of the area and time to be considered, and the incompleteness of the evidence, leave plenty of room for different interpretations about the many phases of the history of the Colorado River. A major purpose of this chapter is to provide a frame of reference into which isolated bits of new evidence can be fitted." In his introduction to the June 2000 Symposium Monograph, Dick Young summed up: "The presence of a well-integrated ancestral upper Colorado River drainage system in portions of Colorado and southern Utah upstream from the Henry Mountains in early Miocene time, as postulated by Charles B. Hunt, cannot be absolutely demonstrated."

Of course, that the upper Colorado River is young, or, if it is not young, that it dried up and sank out of sight west of the Kaibab, are speculations. Yet, to paraphrase Sherlock Holmes, when you have eliminated all the obvious possibilities, something else, however improbable, must be true.

Let us turn from speculation to what we can say with confidence. It is that the ultimate cause of the Grand Canyon is plate tectonics. Thanks to the restless movement of the giant slabs that segment the earth's exterior and interact with each other, the American Southwest contained all the ingredients necessary for extreme canyon cutting: a region high above but laterally near its base level, giving its rivers an unusually steep gradient; water from a high-elevation source a thousand miles away that never dried up; according to the consensus view, knickpoints generated by the opening of the Gulf of California and by repeated faulting in the western Grand Canyon; and, according to Lucchitta, uplift that has continued right to the present day.

⌐_⌐

IN ITS RELENTLESS QUEST TO find and maintain its equilibrium profile, the Colorado River has triumphed over solid rock of every type. Evidently, under the right conditions and given enough time, the physical forces that drive a river inexorably toward that profile are so powerful that they allow a river to overcome each rock and each geologic structure in its path. We could go so far as to say that by maintaining themselves—persisting, as Dutton would have put it, and in a sense he was right after all—the streams of the Colorado Plateau have survived even the worst that plate tectonics could throw at them. In the process, they have had to change almost everything about themselves.

If, as Dutton would have, we probe to find the deepest lessons that the Grand Canyon offers, we discover that they serve to reveal the uniqueness of Planet Earth. There are three such lessons. The first is that plate tectonics continually rearranges the surface of the earth, causing a region to subside at one time and to heave up at another. Plate tectonics affords drainage systems little time for rest or for the passive submission that William Morris Davis's theory envisioned. The second lesson is that plate tectonics requires prodigious amounts of energy. Only a planet that is geologically active—that still has enough internal heat to churn its mantle into giant convecting cells—can supply that energy. In our solar system, Earth is the only rocky planet that retains it internal fires. Other planets may have had some form of plate tectonics early in their history, but not for the last several billion years. No plate tectonics, no Grand Canyon—or anything else that we would recognize.

The third lesson is the power with which geologic time endows running water. Our planet is the only one that we know has liquid water on its surface today, though Mars may have had early in its history. Running water not only carved the Grand Canyon, in Newberry's words, it brought about "a vast system of erosion." But before the sedimentary rocks of the Colorado Plateau could even exist to have a canyon cut through them, water had to produce the sediments that now compose those rocks. As Hutton's great cycle illustrates, such sediments originate when water erodes still-older rocks. As we attempt to trace the process backwards, we soon become lost in an abyss of time, with no vestige of a beginning. Without water and time, sedimentary rock could not exist in the first place. This is what Dutton meant when he wrote, in a section quoted earlier, that "The quantity of material which the agents of erosion deal with is in the long run exactly the same as the quantity dealt with by the agencies of deposition; or, rather, the materials thus spoken of are one and the same. If, then, we would know how great have been the quantities of material removed in any given geological age from the land by erosion, we have only to estimate the mass of the strata deposited in that age. Constrained by this reasoning the mind has no escape from the conclusion that the effects of erosion have been vast." W. M. Davis made the same point when he said, "The earth is necessarily not merely old enough to have allowed the erosion of the canyon in the plateaus, but old enough to have allowed, before the erosion of the canyon was begun, the formation of the huge pile of strata of which the plateaus are composed."

In the few thousand years that the geologists of two centuries ago allowed for earth history—and creationists today allow only 10,000 years, even though written human history extends back for 8,000 years—running water is a puny thing compared to solid rock. Without geologic time, neither sedimentary rocks nor canyons could exist. Deep time gives water dominion over rock, multiplying the meager annual effects of stream erosion innumerable times.

These then are the true lessons of the Colorado River, the Grand

Canyon, and modern geology: our planet and the solar system operate on a vastly different timescale than our human one. Earth is the only planet in our system with fire inside and blue water atop. Earth is unique in this solar system—and perhaps in any.

⌐⌐

A GREAT SCIENTIFIC THEORY not only answers the question it set out to answer, it also leads to unexpected insights and carries deep implications. The theory that large-scale drainage reversal, headward erosion, and stream piracy, possibly aided by lake integration, produced the modern Colorado River meets the test. Contrary to the conception of the pioneers, rivers do not simply grow older, deepening and widening their channels as they lower the landscape to a peneplain. Instead, they engage in a relentless competition, though one that is invisible on a human time scale, as one stream captures another and still others slyly conjoin lakes, only then to drain and destroy them.

In science, and especially in geology, our common sense, our intuition have proven to be fallible guides. True, a few geniuses do seem to have successfully relied on intuition, but the rest of us have not been so able or inspired. Instead, we have to proceed by trial and error, getting a lot wrong and a few things right, slowly advancing science bit by bit. In any case, most of what is important about physics, chemistry, and mathematics, for example, lies well outside our everyday experience, leaving us with no way to apply our common sense to them. Astronomy and geology are temptingly different in that they explain processes that we see going on around us and about which therefore we think we know something. But what people have thought about sky and earth, as often as not, has been dead wrong.

Each morning we see the sun rise and each evening set. Common sense told our ancestors that the sun travels around the earth. Using reason rather than common sense, Copernicus and other astronomers showed that, to the contrary, the earth circles the sun. But this requires our world to hurtle through space at an incomprehensible yet imperceptible speed. Thanks to the invisible force of gravity, everything on earth, including the atmosphere, moves together so that we experience no sensation of the great speed with which we travel. What could make less common sense than gravity?

Without radiometric dating, we would have no way to suspect that *Homo* goes back for several million years and that the earth itself is 4.5 billion years old. Indeed, the passage of billions of years is so far outside our intuition that no one, including geologists, can comprehend what it means. Common sense gives us no reason to suppose that the continents have moved continuously about the globe—what could be less moveable than an entire continent? No one has ever seen a meteorite fall and blast out a crater, yet studies of cratering on the moon and other planets in our neighborhood of space prove that thousands of meteorites, some of them huge, have struck the earth during its long history. One collision wiped out the dinosaurs; it may not have been the only one to cause a mass extinction. To the question of what killed the dinosaurs, the refined intuition of generations of biologists and geologists gave the wrong answer (actually, a hundred different wrong answers). But we ought not be overly critical of our predecessors, for erosion over geologic time has obscured or removed most terrestrial craters. Lacking any strong evidence that meteorite impact affects the earth, and having the false comfort of uniformitarianism, geologists saw no need to appeal to a *deus ex machina* to solve earthly problems.

The Grand Canyon, which at first seems to have an obvious origin, proved hard to understand on second thought and even on the third. J. S. Newberry, the first geologist to see the Canyon, immediately squashed the simple idea that some awful force ripped the earth apart and provided a ready channel for the river. Powell and the other pioneers went on to claim that either the Colorado Plateau had risen underneath an older river or a younger one had descended onto the Plateau by eroding its way down through a sedimentary cover. They were wrong as well. What really happened to create the Grand Canyon was as counterintuitive as deep time, continental drift, and meteorite impact.

So-called common sense leads us astray because it is so hard to appreciate the consequences of geologic time. On a timescale of years, we can see gullies forming and eating their way into a farmer's field.

Within our lifetimes, floods leave rivers hundreds or thousands of feet from their previous channels. But no one has lived long enough to see one stream system work its way upslope and capture another. Yet Gilbert's irrefutable logic and the geologic evidence show that stream capture has happened—and in the Grand Canyon, on a giant scale.

The Colorado River teaches us how rivers evolve not because it is the largest or longest—among U.S. rivers its volume ranks well down the list. Rather, the Colorado is instructive ultimately because, by cramming ridges, plates, subduction zones, and transform faults successively into a single, relatively small region of the earth, plate tectonics forced the river to confront a set of terrains of different ages, structures, altitudes, and slope directions. Had there been no Farallon Plate and no zone to subduct it, all of western North America would have had a completely different geologic history. Thus, for the amazing landscapes of the American Southwest, as for so much else, we can thank plate tectonics.

The consensus among geologists today is that one of the most important lessons of the Grand Canyon is the power of headward erosion. Powell must have been aware of the process, even though he did not mention it in his tripartite classification of streams. Gilbert described headward erosion, but with no suggestion that he thought it capable of carving the lower Colorado River, crossing the Grand Wash Cliffs, and excavating the Grand Canyon all the way to and across the Kaibab. Indeed, his example was the badlands topography of the Dakotas, whose relief is miniscule compared to that of the Grand Canyon.

Though a captured stream has not gone extinct, it has given up its identity to a more energetic one and thus may be analogous to an extinct species. In *Origin of Species*, Charles Darwin called his essential concept "Natural Selection, or Survival of the Fittest." He began by noting the natural variability of organisms that allows breeders of plants and animals to select those that are most useful to humans. Darwin reasoned that this same variability in nature endows some organisms with traits that will increase their chances of survival and procreation. They will leave more offspring and, as the advantageous qualities pass on and multiply in subsequent generations, the organisms that have the most favorable traits will supplant those that do not.

Though Darwin did not know about genes, the mechanism by which organisms bequeath their traits, he deduced that such a mechanism must exist.

The lesson of the Grand Canyon is that rivers and landscapes also evolve through a process in which only the fittest survive. Ivo Lucchitta has been the most vigorous in urging this concept. Just as some traits favor survival of an individual organism and its genes, so certain qualities give a river the energy to cut downward and headward more rapidly, making it more apt to intersect other rivers and survive at their expense. As we have seen, the most important of these qualities is a river's gradient, which derives from its height above its base level. Also important are the amount of water it carries, the amount of silt suspended in its water, the degree to which it runs year-round, the resistance of the rocks over which it runs, whether it contains knickpoint generators, and so on. As with species, the more favorable traits a river possesses, the more likely it is not only to survive, but to capture other streams, if not whole drainages. As the process continues, one landscape—the one with the most energetic streams—replaces another.

WE CAN MAKE A STRONG CASE THAT the more counterintuitive a scientific finding, the more likely it is to have deep implications. Radiometric dating allows us not only to measure the ages of rocks and better understand geologic history, but also to measure the length of geologic time. The surprising discoveries of deep space and deep time then force us to contemplate our place in an incomprehensibly vast and ancient universe. For most of us, the realization that we live for an infinitesimal fraction of the life of the universe, on a planet in a perhaps not atypical solar system, in an average galaxy among billions of galaxies, leads to difficult thinking.

The discovery of continental drift and plate tectonics, which took half a century to find acceptance, brought not only a vastly better understanding of our planet, but the realization that we owe the existence of continents to plate movement. No plate tectonics—no continents—no

Homo sapiens. That in turn makes us consider the kinds of planets that might be eligible for plate activity and which therefore might have intelligent life. Some scientists are convinced that the set of circumstances that led to intelligent life on earth is rare if not unique; in their view, we may be alone in the cosmos.

The serendipitous finding of cosmic levels of the element iridium in the K-T boundary clay by Luis and Walter Alvarez led to the realization that an impacting meteorite killed the dinosaurs and seventy percent of their fellow species. Most geologists objected, a representative one writing, "It is intuitively more satisfying to seek causes from amongst those phenomena which are comparatively familiar to our experience." But the recognition of just how wrong we were about dinosaur extinction brings another difficult realization: without the collision that led to the death of the dinosaurs and made more room for a lucky, hamster-sized mammal, we would not be here. The interest in meteorite impact brought on by the space age and the Alvarezs' discovery in turn led to the understanding that from the very beginning, violence from colliding meteorites has marked our solar system and, among other outcomes, created the moon. Some day—which day we do not know but need to find out—an approaching comet or asteroid will again threaten life on earth.

Conversely, it is often what people's common sense told them about the earth and the solar system that turned out to be the real myth: the earth is only a few thousand years old, is flat, was created just in time for us, circles the sun, contains permanently anchored continents, is immune to extraterrestrial events, and conducts biologic and landscape evolution always at its own uniformitarian pace. Ultimately, what may be most important about astronomy and geology is less their discovery of astounding worlds of space and time than the way in which they demonstrate how the human mind can, but does not necessarily, bring reason to triumph over dogma. Where our ordinary human failings and prejudice lead us astray, reason can, if we let it, set us right. Who cannot fail to be astounded by what science has accomplished since Steno picked up the shark's tooth?

Newberry, Powell, Gilbert, and Dutton believed that by studying the most magnificent of river valleys, they would learn great lessons that would apply everywhere. They were right, but understandably, given

how little the geologists of their day could know, they could not antici-
pate the great lessons that derived from the Grand Canyon. The
pioneers were able immediately to jettison the jejune notion that valleys
come first and rivers second. But then, unaware of the potential of headward erosion, Powell and Dutton embraced a false god: the Law of the
Persistence of Rivers. Not until the late twentieth century and the
advent of age dating and plate tectonics did the true import of the
Grand Canyon emerge: that plate movement causes whole landscapes to
fall and others to rise, empowering rivers to remarkable feats. Nor did
anyone suspect that headward erosion was such a powerful agency, forcing rivers to engage in a fierce fight for slope and territory and allowing
only the fittest to survive for geologic epochs.

In the period between J. W. Powell's first voyage and the June 2000
Symposium, geology experienced three scientific revolutions: recognition that the earth is billions of years old; that continents drift, caught
on the giant conveyor belt of plate techtonics; and that extraterrestrial
events have had a major effect on earth and biologic history. In a sense,
these revolutions reflect a shift from the view of the earth as young,
fixed, and uniformitarian, to one of deep time, change, and now and
again, catastrophe. Naturally, the Major and his assistants and contemporaries could know nothing of these future paradigm shifts. What they
could and did do is make careful observations that would stand long
after their particular interpretations of what they saw had been
superceded. That is truly what it means to be a geologist. Those who
take the time to observe deeply and carefully also are revolutionaries.

The laws of physics require rivers to always seek their equilibrium
profiles. Remarkably, in spite of solid rock and seemingly, even in spite of
plate tectonics, they succeed. If this is not apparent in a great river like
the Mississippi, wait until future plate movements rearrange North America. Louisiana may no longer be lower than Minnesota. Some now third-
rate tributary may approach the Mississippi from some then-lower territory. As in a Shakespearean tragedy, the Father of Waters then will have to
fight his relatives for survival. The wait will be long; it will take time. But
the lesson from the Colorado Plateau, and indeed from all of modern
geology, is that of time, the earth has more than we can conceive.

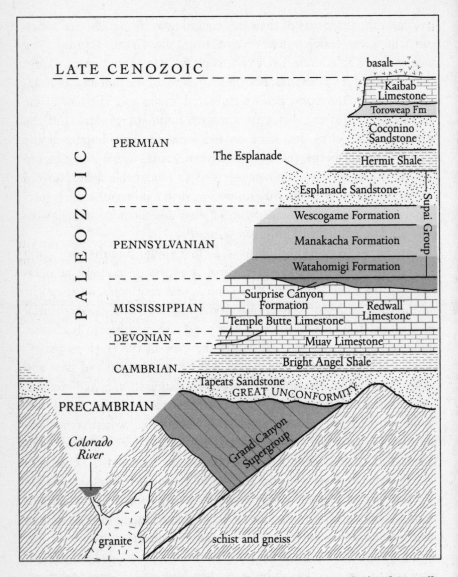

Figure 16. *Geologic cross-section of the Grand Canyon, after Potochnik and Reynolds.*

Key Terms and Places

Antecedent stream: One which established its course before the terrain rose beneath it, causing the river to cut a canyon and making the stream older than the structures it crosses. J. W. Powell invented the name.

Antelope Canyon: One of many narrow slot canyons that enter the Colorado River. At Antelope Canyon, near Lake Powell, 11 people lost their lives in a flash flood in 1997.

Anticline: Upfold or arch of stratified rock in which the layers bend downward in opposite directions from the crest, in the shape of an upside down U.

Aquarius Plateau: The easternmost of the High Plateaus of Utah, near Capitol Reef National Park. Almon Thompson, Powell's brother-in-law, named the plateau Aquarius after the water bearer of Greek mythology, in honor of its many lakes.

Argon-argon dating: The modern, more reliable successor to the older potassium-argon method of dating rocks.

Base level: The lowest level to which a stream can erode. The elevation of its junction either with a larger stream or sea level.

Basin and Range Province: The region of block-faulted mountains with intervening basins that lies to the west of the Colorado Plateau, comprising most of Nevada, southwest Arizona, and southeastern California. The name is attributed to G. K. Gilbert.

Bidahochi Formation: The thick sequence of Pliocene sedimentary rocks found in Northern Arizona, southeast of the Grand Canyon, and believed by geologists to have accumulated in now-vanished Hopi Lake.

Book Cliffs: An east-west escarpment south of the Uinta Range at the foot of Gray Canyon on the Green River.

Boulder Dam: The former name of Hoover Dam, which impounds Lake Mead in southest Nevada.

Bouse Formation: Early Pliocene sedimentary deposits found in basins between Lake Mead and the Gulf of California. Geologists believed it to have formed in a delta, marking the first appearance of the Colorado River below the Grand Canyon.

Brown's Park: An elongated valley northeast of the Uinta Range, on the Green River, where Powell and his men rested.

Cataract Creek: A long tributary to the Colorado in the western Grand Canyon. Farther downstream its name changes to Havasu Creek.

Catastrophism: The doctrine that geologic change happens suddenly and through processes we cannot observe. The countervailing theory is uniformitarianism.

Coconino Plateau: The table-land which makes up the South Rim of the Grand Canyon. The Kaibab Plateau makes up the North Rim, but stands one thousand feet higher than the Coconino Plateau. Coconino means "little water" in the Havasupai Indian language.

Colorado Plateau: The large region of the American Southwest characterized by the uplifted block of largely flat-lying Paleozoic and Mesozoic rocks and by vertical cliffs and and deeply-incised canyons. Pyne attributes the naming of the plateau to Gilbert.

Consequent stream: One that flows over a fresh, uneroded landscape. Powell coined the term.

Contraction model: The theory that the earth has been cooling since it first formed and that the resulting contraction is the engine that creates mountains and drives geologic processes. Analogous to the wrinkled skin of an apple. Scientists rejected the model when they discovered isostasy and radioactivity.

Crossing of the Fathers (El Vado de los Padres; Ute Ford): The site above Lee's Ferry where Fathers Dominguez and Escalante forded the Colorado while returning from their failed attempt to reach Monterey, California, from Santa Fe; now under Lake Powell.

Desolation Canyon: The canyon of the Green River just north of Gray Canyon, both located north of the present-day town of Green River, Utah. Powell named them both.

Diamond Creek: The tributary entering the Colorado at river mile 226, just upstream from dry Peach Springs Canyon. Lieutenant Ives named the creek.

Dynamic equilibrium: The theory that rivers, and many other natural systems, always adjust the factors that affect them so as to do minimum work. Thus, when humans intervene in a stream, as by damming it, the stream reacts in such a way as to offset the intervention. The term is attributed to Gilbert.

Eocene Epoch: The fourth oldest of the epochs of the Tertiary period of the Mesozoic era, 55.5–33.7 million years.

Eohippus: A genus of small, primitive four-toed horses from the Lower Eocene of the western United States. Important in supporting Darwin's theory of evolution.

Erratic: A large rock transported by a glacier from its original outcrop miles away.

Esplanade: A broad bench below the rim of the western Grand Canyon. Dutton gave it its name.

Farallon Plate: The tectonic plate that disappeared down the subduction zone that once lay off southwestern North America. Its subduction was instrumental in creating the topography of the western United States.

Flaming Gorge: The canyon of the Green River north of the Uinta Range and named by Powell for its striking red rock. Now submerged under the reservoir of the same name.

Flood geology: The belief that the Bible is the literal truth, that the earth is only a few thousand years old, and the great Flood of Noah is its major geological event.

Geographical cycle: The theory of William Morris Davis that landscapes pass through youth, maturity, and old age. The concept of dynamic equilibrium long ago replaced it.

Geomorphology: The science of the evolution of landscapes. Powell, Dutton, Gilbert, and Davis collectively invented it.

Glen Canyon: The beautiful canyon upstream from Marble Canyon, now submerged under Lake Powell. Powell named the canyon for its "curious ensemble of wonderful features—carved walls, royal arches, glens, alcove gulches, mounds, and monuments."

Graded stream: The term invented by Gilbert to denote a stream in dynamic equilibrium, in which the stream balances all the factors that affect it so as to do minimum work.

Grand Canyon: Powell instructed Dellenbaugh to use this name instead of Big Cañon, as the Spaniards had called it.

Grand Canyon Supergroup: The set of tilted Precambrian beds exposed at the bottom of the Grand Canyon.

Grand River: The original name of the Colorado River upstream from its junction with the Green River. The Colorado Legislature renamed the Grand River the Colorado River in 1921.

Grand Staircase: The succession of Mesozoic cliffs and valleys that descend from the High Plateaus of Utah toward the Grand Canyon. The term is often attributed to Dutton but Charles Keyes coined it in 1924.

Grand Wash Cliffs: The western border of the Grand Canyon and the site of a large fault; the boundary between the Colorado Plateau and the Basin and Range Province.

Gray Canyon: Cut by the Green River; near the present-day town of Green River, Utah. Immediately south of Desolation Canyon.

Great Unconformity: The geologic boundary between the oldest Paleozoic formation in the Grand Canyon, the Tapeats Sandstone, and the tilted Precambrian rocks beneath them. At this level, more than one billion years worth of rocks are missing.

Green River: The longer of the two streams that meet to form the original Colorado River, the other being the Grand River. Since the Green River is the longer of two major branches, it is the true headwaters of the Colorado River.

Gulf of California (Sea of Cortez): Body of saltwater between Baja California and the mainland, opened by plate movement about five million years ago.

Havasu Creek: A stream that enters the Colorado River at river mile 157; it is one of few large tributaries to the Grand Canyon. Upstream it is called Cataract Creek and is home to the Havasupai.

Havasupai: The tribe of Native Americans that previously had lived for hundreds of years in isolation along Havasu Creek.

Headward erosion: The lowering and lengthening of a streambed by upstream retreat of a knickpoint. The process leads one drainage to capture another.

Henry Mountains: A small range north of the eastern end of Lake Powell, made famous among geologists by the seminal studies of Gilbert.

High Plateaus: The group of generally flat-topped table-lands in central and eastern Utah that reach as high as 12,000 feet.

Hoover Dam: Impounds the Lower Colorado River and Lake Mead in southeast Nevada. Hoover was the first large dam built by the Bureau of Reclamation; it was completed in 1936.

Hopi Lake: Hypothetical Miocene-Pliocene lake southeast of the Grand Canyon, where the Little Colorado River now flows.

Hualapai Limestone: A 1,000-foot-thick bed of limestone at the top of the Muddy Creek Formation, just west of the mouth of the Grand Canyon. The limestone is the youngest rock unit to pre-date the Grand Canyon.

Hualapai Plateau: A table-land on the south rim of the western Grand Canyon that is home to the Hualapai Tribe. The plateau includes the town of Peach Springs, Arizona, and Peach Springs Canyon, as well as Diamond Creek. Hualapai is the Indian word for Pine Tree Folk.

Hurricane Fault: A large fault that together with the Toroweap Fault crosses the Colorado River in the western Grand Canyon. Both extend for scores of miles north and south of the Grand Canyon.

Imperial Formation: The sedimentary unit that is just younger than, and found just farther south than, the Bouse Formation. Both are believed to represent delta deposits from the Colorado River from when it first entered the Gulf of California.

Isostasy: The theory that blocks of the earth's upper surface float in gravitational balance according to their dimensions and density. The theory requires that some zone of the earth's interior be able to flow. Isostasy was critical to the development of the theory of continental drift. Proposed by Dutton.

Kaibab: An Indian word meaning "mountain-lying-down," according to J. W. Powell. Applied to the Kaibab Plateau on the north rim in the eastern Grand Canyon, which at 9,000 feet is the highest region to adjoin the canyon. Geologists use Kaibab Uplift, Kaibab Unwarp, and Kaibab Arch interchangeably.

Kaiparowits Plateau: A table-land running northwest-southeast just below the Escalante River in southeast Utah.

Kanab: The word comes from the Paiute term for willow. Kanab refers to the town in southern Utah just north of the Grand Canyon; to the creek and canyon of the same name that enter the Colorado at river mile 143.5, and to one of four large plateaus on the north rim of the Grand Canyon.

Knickpoint: An abrupt break in the slope of a streambed that causes the stream to speed up and erode more forcefully; a waterfall is the extreme example. A knickpoint retreats upstream and leaves a deeper channel in its wake.

Lake Bonneville: The much larger ancestor of the present Great Salt Lake and other Utah lakes; studied by Gilbert.

Lake Mead: The large body of water impounded by Hoover Dam, drowning the Colorado River valley west of the mouth of the Grand Canyon.

Land bridges: Hypothetical isthmuses that supposedly explain how identical organisms could have lived on continents separated by thousands of miles of ocean water. Invented so that geologists did not have to accept the theory of continental drift.

Laramide Orogeny: The series of mountain-building events that created the Rocky Mountains and the western Cordillera from Canada to northern Mexico. Originally geologists thought the orogeny marked the Cretaceous-Tertiary boundary, which they later dated to 65 million years ago. Though most of the mountain uplift did take place then, subsequent work showed that the orogeny actually comprises several discrete pulses that which may have lasted into the Oligocene epoch, 25–35 million years ago. Plate tectonics, probably the continued thrusting of the Farallon Plate under the American Plate, caused the Laramide Orogeny.

Lee's Ferry: A relatively shallow-water site at the north end of Marble Canyon and the mouth of the Paria River named after Mormon John D. Lee, who operated a ferry. Powell's men rested at Lee's Ferry on their second voyage. Authorities tried and executed Lee in 1877 for participating in the murder of 120 emigrants at Mountain Meadows.

Little Colorado River: The large tributary that enters the Colorado River at river mile 61.5.

Lodore Canyon: The canyon cut by the Green River through the eastern Uinta Range, the first white water Powell and his men met. Lodore is named for a poem by Robert Southey.

Marble Canyon: A relatively straight stretch of the Grand Canyon from Lee's Ferry to the entry of the Little Colorado River at mile 61.5. Powell named it Marble Canyon for the prominent Redwall Limestone.

Markágunt Plateau: The southwesternmost of the High Plateaus of Utah, site of present-day Zion Canyon National Park.

Miocene Epoch: The second oldest of the epochs of the Tertiary period of the Mesozoic era, about 23.8–5.3 million years.

Mogollon Highlands: The upturned southern rim of the Colorado Plateau, extending across central Arizona.

Mojave River: A stream in Southern California which Norman Meek believes was formed by the integration of lakes.

Monocline: A stepfold in which roughly horizontal rock beds connect an inclined section.

Muddy Creek Formation: A thick sequence of Miocene sedimentary rocks deposited in interior basins just west of the mouth of the Grand Canyon. The Colorado River could not have flowed through the region until after the Muddy Creek sediments accumulated.

Oligocene Epoch: The third oldest of the five epochs of the Tertiary period of the Mesozoic era, about 33.7–23.8 million years.

Neptunism: The rival to the theory of uniformitarianism of James Hutton, associated with Abraham Gottlob Werner, who claimed that all rocks had precipitated on the floor of a universal ocean.

Páhvant Plateau: The northwestern-most of the High Plateaus of Utah.

Paleocene Epoch: Oldest of the epochs of the Tertiary period of the Mesozoic era, about 65–56.5 million years.

Paria Plateau: The table-land on the west side of the Paria River north of Lee's Ferry.

Paria River: The stream that enters Marble Canyon at Lee's Ferry. The name comes from a Ute word that means "elk water" or "dirty water."

Paunságunt Plateau: The southeasternmost of the High Plateaus of Utah. It overlooks the Grand Staircase and, in the distance, the Kaibab Plateau and the Grand Canyon.

Peach Springs Canyon: A dry canyon that slopes from southwest to northeast to approach the Colorado River near the mouth of Diamond Creek; visited by Garcés in 1776. The name comes from the peaches that the Spaniards gave the Hualapai Tribe.

Peach Springs Lake: A body of water hypothesized by geologist Charles B. Hunt to have existed on the Hualapai Plateau. Hunt conjectured that the lake had drained through subterranean pipes which exited below the Grand Wash Cliffs to fill Hualapai Lake.

Peach Springs Tuff: An 18-million-year-old volcanic rock found in Peach Springs Canyon.

Peneplain: A nearly flat plain of little relief; the last stage in long-continued erosion. Davis coined the term, which means "almost a plain."

Pleistocene Epoch: The oldest of the epochs of the Quaternary period of the Cenozoic era, about 1.8 million years to about 10,000 years.

Pliocene Epoch: The youngest of the epochs of the Tertiary period of the Mesozoic era, about 5.3–1.8 million years.

Plate tectonics: The modern theory in which thin, rigid slabs move over a fluid layer. The collision, separation, and subduction of plates accounts for most large-scale features of the earth's surface, including the opening of the Gulf of California.

Potassium-argon dating: An older method of absolute age dating based on the decay of potassium-40, with a half-life of about 1.3 billion years, to argon-40. In the past few decades, the more accurate argon-argon method has replaced it.

Radiometric age dating: The method of measuring rock ages based on knowing the amount and half-life of a parent isotope, plus the amount of its daughter isotope.

Rim gravels: The outcroppings of rounded pebbles and cobbles found on the Hualapai Plateau, far from their source to the south on the Mogollon Highlands. The gravels show that ancient drainage in northern Arizona flowed from south to north, at right angles to its current direction.

River miles: The traditional way to locate points within the Grand Canyon, measured as miles downstream along the river, starting from Lee's Ferry.

Roan Cliffs: The long staircase-like escarpment through which the Green River has cut Gray Canyon, north of present-day Green River, Utah, and just north of the Book Cliffs.

Salton Sea: A saline lake in southern California created in 1905–1906 when the Colorado River broke through diversion dams and filled this remnant of a prehistoric lake. Irrigation waters now supply it.

San Andreas fault: The 1,300 km fault running from the Gulf of California to north of San Francisco. It forms the boundary between the American and Pacific Plates.

Sea of Cortez: The alternative name for the Gulf of California, separating mainland Mexico and Baja California.

Sevier Plateau: One of the High Plateaus of Utah, north of present-day Richfield, and site of Mount Dutton.

Shivwits Plateau: The westernmost of the plateaus that make up the north rim of the Grand Canyon; named for the Shivwits Tribe.

Slot canyon: A canyon that is many times narrower than deep. Such canyons are typical of the tributaries that enter the Colorado River in the Grand Canyon; they are especially well-developed in beautiful Zion Canyon National Park.

Steno's Law: The principle that layers of rock accumulate horizontally, with the oldest at the bottom and the youngest at the top.

Stream piracy: A process by which headward erosion of a more-energetic stream causes it to intersect and capture the waters of a less energetic one.

Super(im)posed stream: A stream which erosion has lowered through a cover of sedimentary rock onto a buried older structure below. J. W. Powell coined the term.

Syncline: A downfold in layered rock in the form of a U.

Toroweap: The companion large fault to the Hurricane Fault. Also a lookout point on the North Rim. The word is the Paiute term for "dry or barren valley."

Transform fault: One of the possible boundaries between two tectonic plates, in which the two slide past each other. The San Andreas is one of the few that appear on land.

Tushar Plateau: A west-central member of the High Plateaus of Utah.

Uinkaret Plateau: One of the large plateaus on the North Rim of the Grand Canyon, east of the Shivwits Plateau and west of the Kanab Plateau. Uinkaret is a Piute word meaning "where the pines grow."

Uinta Range: Lofty mountains along the border between Utah and Wyoming. In contrast to most other ranges in the western hemisphere, which trend north-south, the Uintas run east-west. Kings Peak at 13,528 feet is the highest point in Utah.

Uniformitarianism: The theory proposed by James Hutton and developed by Charles Lyell which holds that "the present is the key to the past": we can deduce everything that is important about geologic history by observing processes at work today. Over time, the theory became proscriptive: nothing that geologists cannot observe today, like continental drift and meteorite impact, can have been important.

Virgin River: The stream that flows through southwestern Utah and Zion National Park to enter Lake Mead. In J. W. Powell's day, the Spanish spelling Virgen was often used. The Virgin River was the destination of his succesful first river expedition.

Wasatch Range: Lofty mountains east of Salt Lake City.

References

Chapter 1. Six Feet

"He declared it 'altogether valueless'": Ives, J. C. (1861). *Report Upon the Colorado River of the West, Explored in 1857 and 1858 by Lieutenant Joseph C. Ives . . . Under the Direction of the Office of Explorations and Surveys.* Washington, D.C., U.S. Government Printing Office.

"The Colorado plateau is to the geologist a paradise": Newberry, J. S., J. N. Macomb, et al. (1876). *Report of the Exploring Expedition from Santa Fé, New Mexico, to the Junction of the Grand and Green Rivers of the Great Colorado of the West in 1859.* Washington, D.C., Engineer Dept. U.S. Army: G.P.O. 54

"The grand cañon of the Colorado will give the best geological section": Worster, D. (2002). *A River Running West: The Life of John Wesley Powell.* New York, Oxford University Press. 127

"The Plateau province offers valuable matter": Gilbert, G. K. (1876). "The Colorado Plateau Province as a Field for Geological Study." *American Journal of Science, Third Series* XII: 16-24, 85-103. 18

"It would be difficult to find anywhere else in the world": Dutton, C. E. (1882). *Tertiary History of the Grand Cañon District.* Reston, VA, U.S. Geological Survey. 92

"For every difficult question": Attributed to G. B. Shaw

Chapter 2. Water Catch 'Em

"An Odd Fleet": These authors tell the story of Powell's famous launch: Darrah, W. C. (1951). *Powell of the Colorado.* Princeton, Princeton University Press. Dolnick, E. (2001). *Down the Great Unknown: John Wesley Powell's 1869 Journey of*

Discovery and Tragedy Through the Grand Canyon. New York, HarperCollins. Stegner, W. E. (1992). *Beyond the Hundredth Meridian: John Wesley Powell and the Second Opening of the West.* New York, Penguin Books. Worster, D. (2002). *A River Running West: The Life of John Wesley Powell.* New York, Oxford University Press.

"I suppose you never herd off him": Rabbitt, M. C. (1969). "John Wesley Powell, his life & times." *Geotimes* 14(5): 10-12

"It was not a government expedition": Stegner introduction to Penguin edition Powell, J. W. (1875). *Exploration of the Colorado River of the West and Its Tributaries.* Washington, D.C., U.S. Government Printing Office. vii

"To find out": Stegner introduction to Penguin edition Ibid. viii

"The Grand Canyon of the Colorado": Pyne, S. J. (1999). *How the Canyon Became Grand: A Short History.* New York, Penguin Books. 57

"There are great descents yet to be made": Powell, J. W. (1875). *Exploration of the Colorado River of the West and Its Tributaries.* Washington, D.C., U.S. Government Printing Office. 62

"'The rocks,' he said": Ibid. 16-17

"Between Green River, Wyoming and the mouth of the Uinta": Stegner, W. E. (1992). *Beyond the Hundredth Meridian: John Wesley Powell and the Second Opening of the West.* New York, Penguin Books. 43

"No one heard a word": Ibid. 54

"On the 7th or 8th day of May": (Saturday, July 3, 1869). "Fearful Disaster." *Chicago Tribune.* Chicago. www.pbs.org

"Risdon even earned an audience": Dolnick, E. (2001). *Down the Great Unknown: John Wesley Powell's 1869 Journey of Discovery and Tragedy Through the Grand Canyon.* New York, HarperCollins. 132

"Moreover, she had received a letter": Ibid. 131, 137

"the report of my death was an exaggeration": (1979). *The Oxford Dictionary of Quotations.* Oxford, Oxford University Press. 528

"'For two hours,' Risdon reported,": Dolnick, E. (2001). *Down the Great Unknown: John Wesley Powell's 1869 Journey of Discovery and Tragedy Through the Grand Canyon.* New York, HarperCollins. 133. Worster, D. (2002). *A River Running West: The Life of John Wesley Powell.* New York, Oxford University Press. 174

"What better proof": Dolnick, E. (2001). *Down the Great Unknown: John Wesley Powell's 1869 Journey of Discovery and Tragedy Through the Grand Canyon.* New York, HarperCollins. 133

"Together they searched the woods": Dellenbaugh, F. S. (1998). *The Romance of the Colorado River: The Story of Its Discovery in 1540, . . .With Special Reference to the*

Voyages of Powell Through the Line of the Great Canyons. Mineola, NY, Dover Publications. 374

"While a student, Hayden too became enthusiastic": Foster, M. (1994). *Strange Genius: The Life of Ferdinand Vandeveer Hayden.* Niwot, CO, Roberts Rinehart Publishers. 20

"In 1858, Powell returned to Wheaton": Dellenbaugh, F. S. (1998). *The Romance of the Colorado River: The Story of Its Discovery in 1540, . . . With Special Reference to the Voyages of Powell Through the Line of the Great Canyons.* Mineola, NY, Dover Publications. 374-375

"After all, it was Powell": Worster, D. (2002). *A River Running West: The Life of John Wesley Powell.* New York, Oxford University Press. 124-125

"The exploration was not made for adventure": Powell, J. W. (1997). *The Exploration of the Colorado River and Its Canyons.* New York, Penguin. Preface

"national celebrity": Pyne in introduction to Dutton, C. E. (2001). *Tertiary History of the Grand Cañon District.* Tucson, University of Arizona Press. xv

"glittering public hero": Stegner in introduction to Ibid. vi

"the most celebrated adventurer since Lewis and Clark": Reisner, M. (1987). *Cadillac Desert: The American West and Its Disappearing Water.* New York, Penguin Books. 51

"Enters the range by a flaring, brilliant, red gorge": Powell, J. W. (1875). *Exploration of the Colorado River of the West and Its Tributaries.* Washington, D.C., U.S. Government Printing Office. 11

"One pioneer had said": Dolnick, E. (2001). *Down the Great Unknown: John Wesley Powell's 1869 Journey of Discovery and Tragedy Through the Grand Canyon.* New York, HarperCollins. 56

"An unromantic member of the crew": Ibid. 56

"The Gate of Lodore hinges inward": Ibid. 58

"Powell had used the texts of Dana": Worster, D. (2002). *A River Running West: The Life of John Wesley Powell.* New York, Oxford University Press. 116

Chapter 3. How Old Is a River?

"Not only are rivers important": McCully, P. (1996). *Silenced Rivers: The Ecology and Politics of Large Dams.* London; Atlantic Highlands, NJ, Zed Books.

"Yet at the average rate": "River." Encyclopædia Britannica. 2005. Encyclopædia Britannica Premium Service. 13 Jan. 2005. www.britannica.com

"The oldest hominid fossils": McCully, P. (1996). *Silenced Rivers: The Ecology and Politics of Large Dams.* London; Atlantic Highlands, NJ, Zed Books. 9

"The first archeological evidence": Ibid. 9

"Throughout history": "River." Encyclopædia Britannica. 2005. Encyclopædia Britannica Premium Service. 13 Jan. 2005. www.britannica.com

"A Danish anatomist named Niels Stensen": www.ucmp.berkeley.edu

"Werner not only stayed home": "Werner, Abraham Gottlob." Encyclopædia Britannica. 2005. Encyclopædia Britannica Premium Service. 13 Jan. 2005. www.britannica.com

"John F. Kennedy once remarked": Kennedy, J. F. (1962). *Bartlett's Familiar Quotations*. Little, Brown; Boston, Toronto, London. 741

"The prose may appear antiquated": Hallam, A. (1989). *Great Geological Controversies*. Oxford; New York, Oxford University Press. 31

"In examining things present": Hutton, J. (1788). *Transactions of the Royal Society of Edinburgh* I. 217

"It is, however, where rivers issue through narrow defiles": Playfair, J. (1964). "Illustrations of the Huttonian Theory of the Earth"; a facsimile reprint. New York, Dover Publications. 104

"Every river appears to consist of a main trunk": Ibid. 102

"It is there, says Dr. Hutton": Ibid. 351

"But according to his son": Hallam, A. (1989). *Great Geological Controversies*. Oxford; New York, Oxford University Press. 62

"Buckland claimed to have eaten": Ibid. 62

"In Buckland's 1820 inaugural lecture": Ibid. 42

"Faulting, or some other catastrophic event": Buckland, R. W. (1822). "On the Excavation of Valleys by Diluvial Acton." *Transactions of the Royal Society of London* Series 2, I: 95-102. 96

"Buckland saw removal of the rock": Ibid. 96

"It is not easy to imagine": Ibid. 97

"Even after Agassiz showed": Bolles, E. B. (1999). *The Ice Finders: How a Poet, a Professor, and a Politician Discovered the Ice Age*. Washington, D.C., Counterpoint. 234-235

"It is probable that few great valleys": Attributed to Lyell by Thornbury, W. D. (1954). *Principles of Geomorphology*. New York, John Wiley. 7. Lyell restated this same reservation in several other places, for example in the 11th Edition, Chapter IV, p. 77. Lyell, C. (1872). *Principles of Geology; or, The modern Changes of the Earth and Its Inhabitants Considered as Illustrative of Geology*. London, J. Murray.

"In 1877 he attacked uniformitarianism": King, C. (1877). "Catastrophism and evolution." *The American Naturalist*: 449-470

"The uniformitarians needed billions": Worster, D. (2002). *A River Running West: The Life of John Wesley Powell*. New York, Oxford University Press. 315

"Geikie would soon write": Pyne, S. J. (1980). *Grove Karl Gilbert, a Great Engine of Research*. Austin, University of Texas Press. 53

Chapter 4. The Saw That Cut the Mountain

"The drawing of the Gate of Lodore": Powell, J. W. (1875). *Exploration of the Colorado River of the West and Its Tributaries*. Washington, D.C., U.S. Government Printing Office. 153

"In 1875, in response to a request": Ibid.

"squidlike tendency to retreat, squirting ink": Stegner, W. E. (1992). *Beyond the Hundredth Meridian: John Wesley Powell and the Second Opening of the West*. New York, Penguin Books. 52

"the men of the second party": Dellenbaugh, F. S. (1998). *The Romance of the Colorado River: The Story of Its Discovery in 1540, . . . With Special Reference to the Voyages of Powell Through the Line of the Great Canyons*. Mineola, NY, Dover Publications. vi

"Many years have passed since the exploration": Powell, J. W. (1997). *The Exploration of the Colorado River and Its Canyons*. New York, Penguin.

"Stegner thought so": Stegner, W. E. (1992). *Beyond the Hundredth Meridian: John Wesley Powell and the Second Opening of the West*. New York, Penguin Books. 137

"Or, as Worster points out": Worster, D. (2002). *A River Running West: The Life of John Wesley Powell*. New York, Oxford University Press. 257

"To a person studying the physical geography": Powell, J. W. (1875). *Exploration of the Colorado River of the West and Its Tributaries*. Washington, D.C., U.S. Government Printing Office. 152

"Meeting a softer bed": Ibid. 164

"From the foot of Red Canyon": Ibid. 157

"The first explanation suggested": Ibid. 152

"The contracting or shriveling of the earth": Ibid. 153

"The upheaval was not marked": Ibid. 154

"Another illustration of the gradual": Foster, M. (1994). *Strange Genius: The Life of Ferdinand Vandeveer Hayden*. Niwot, CO, Roberts Rinehart Publishers. 113

"Though the entire region has been folded": Powell, J. W. (1875). *Exploration of the Colorado River of the West and Its Tributaries*. Washington, D.C., U.S. Government Printing Office. 198

"Powell biographer Donald Worster says this episode took place": Worster, D. (2002). *A River Running West: The Life of John Wesley Powell*. New York, Oxford University Press. 170

"One of Powell's men": Ibid. 178

"One of the most accessible": www.climb-utah.com/Powell/flash_antelope.htm

"He understood that base level": Powell, J. W. (1875). *Exploration of the Colorado River of the West and Its Tributaries*. Washington, D.C., U.S. Government Printing Office. 203

"Thus ever the land and sea are changing": Ibid. 214

"Bring me men to match my mountains": Foss, Sam Walter. *The Coming American*. www.giga-usa.com

Chapter 5. Seven Cities of Gold

"half the natives died from a disease": The West Film Project (2001). www.pbs.org

"dumbfounded at the sight of me": Ibid.

"powerful villages, four and five stories high": Farish, T. E. (1915). *History of Arizona*. www.southwest.library.arizona.edu

"streets lined with goldsmith shops": www.over-land.com

"When they saw the first village": The West Film Project (2001). www.pbs.org

"He was well received": The West Film Project (2001). www.pbs.org

"the horsemen gave chase": www.xmission.com

"a very brutish people": The West Film Project (2001). www.pbs.org

"I remained twenty-five days": The West Film Project (2001). www.pbs.org

"Why immigrate to Kansas?": Reisner, M. (1987). *Cadillac Desert: The American West and Its Disappearing Water*. New York, Penguin Books. 40

"they garroted him": The West Film Project (2001). www.pbs.org

"Even so, Coronado did leave": Reisner, M. (1987). *Cadillac Desert: the American West and Its Disappearing Water*. New York, Penguin Books. 17

"We were deeper": Ives, J. C. (1861). *Report Upon the Colorado River of the West, Explored in 1857 and 1858 by Lieutenant Joseph C. Ives . . . Under the Direction of the Office of Explorations and Surveys*. Washington, D.C., U.S. Government Printing Office. 107-108

"Whether it would support his weight": Ibid.

"even across terrain that offered not the slightest impediment": Pyne, S. J. (1999). *How the Canyon Became Grand: A Short History.* New York, Penguin Books. 24-25

"was unable to spare a boat": www.nalanda.nitc.ac.in

"It [the Grand Canyon] looks like the Gates of Hell": Ives, J. C. (1861). *Report Upon the Colorado River of the West, Explored in 1857 and 1858 by Lieutenant Joseph C. Ives . . . Under the Direction of the Office of Explorations and Surveys.* Washington, D.C., U.S. Government Printing Office. 110

"politics and money": Lago, Don (2001). James White. *Boatman's Quarterly Review.* 14. www.gcrg.org

"valueless to the agriculturalist": Newberry, J. S., J. N. Macomb, et al. (1876). *Report of the Exploring Expedition from Santa Fé, New Mexico, to the Junction of the Grand and Green Rivers of the Great Colorado of the West in 1859.* Washington, D.C., Engineer Dept. U.S. Army: G.P.O. 54

"the most splendid exposure": Ives, J. C. (1861). *Report Upon the Colorado River of the West, Explored in 1857 and 1858 by Lieutenant Joseph C. Ives . . . Under the Direction of the Office of Explorations and Surveys.* Washington, D.C., U.S. Government Printing Office. 101

"The peculiar topographical features": Newberry, J. S. (1861). Geological Report. *Report Upon the Colorado River of the West, Explored in 1857 and 1858 by Joseph C. Ives.* Washington, D.C., U.S. Government Printing Office, U.S. Army, Topographical Corps of Engineers: B45

"The opposite sides of the deepest chasm": Ibid. B46

"was only a subtext to the larger debate": Pyne, S. J. (1999). *How the Canyon Became Grand: A Short History.* New York, Penguin Books. 49

Chapter 6. America's Greatest Geologists

"which provided and will still provide": Stegner, W. E. (1992). *Beyond the Hundredth Meridian: John Wesley Powell and the Second Opening of the West.* New York, Penguin Books. 157. Ward's will still provide at www.wardsci.com.

"Newberry's old maps, reports, and brilliant hunches": Pyne, S. J. (1980). *Grove Karl Gilbert, a Great Engine of Research.* Austin, University of Texas Press. 37

"the exploration of the Colorado River may now be considered complete": Wheeler, C. G. M. (1889). *Geographical Report; U.S. Geographical Survey West of the 100th Meridian.* Washington, D.C., U.S. Government Printing Office. 170

"One of the reasons": Pyne, S. J. (1999). *How the Canyon Became Grand: A Short History*. New York, Penguin Books. 66

"stand without a known rival": Ibid. 66

"Wheeler's decision to traverse Death Valley": Ibid. 65

"Breeze": Pyne in the introduction to Dutton, C. E. (2001). *Tertiary History of the Grand Cañon District*. Tucson, University of Arizona Press. xv

"They chose Grove Karl Gilbert": Pyne, S. J. (1999). *How the Canyon Became Grand: A Short History*. New York, Penguin Books. 68

"Since Gilbert never took a professorship": Pyne, S. J. (1980). *Grove Karl Gilbert, a Great Engine of Research*. Austin, University of Texas Press. 21

"Gilbert was to measure the landscape": Ibid. 81

"translated river silt and sand into equations": Pyne, S. J. (1999). *How the Canyon Became Grand: A Short History*. New York, Penguin Books. 65

"the closest thing to a saint": Hoyt, W. G. (1987). *Coon Mountain Controversies: Meteor Crater and the Development of Impact Theory*. Tucson, University of Arizona Press. 37

"Great Engine of Research": Pyne, S. J. (1980). *Grove Karl Gilbert, a Great Engine of Research*. Austin, University of Texas Press.

"Powell's better half": Pyne, S. J. (1999). *How the Canyon Became Grand: A Short History*. New York, Penguin Books. 67

"Was extremely fertile in ideas": Gilbert, G. K. (1902). "John Wesley Powell." *Science*: 561-567. 563-564

"America's greatest contribution to scientific philosophy": Worster, D. (2002). *A River Running West: The Life of John Wesley Powell*. New York, Oxford University Press. 448

"In his eulogy": Gilbert, G. K. (1902). "John Wesley Powell." *Science*: 561-567. 567

"To the nation he is known as an intrepid explorer": Ibid. 567

"He was a great man": Rabbitt, M. C. (1969). "John Wesley Powell, his life & times." *Geotimes* 14(5): 10-12. 12

"Together they read the classics": Worster, D. (2002). *A River Running West: The Life of John Wesley Powell*. New York, Oxford University Press. 29

"botanist, geologist, zoologist, ethnologist, archaeologist, historian, philosopher": Stegner, W. E. (1992). *Beyond the Hundredth Meridian: John Wesley Powell and the Second Opening of the West*. New York, Penguin Books. 14

"His interests were by and large Crookham's interests": Ibid. 14

"This family obligation": Bourgeois, J. (1998). "Rock stars; model survey geologist: G. K. Gilbert." *GSA Today* 8(2): 16-17

"He earned more credits in mathematics": Pyne, S. J. (1980). *Grove Karl Gilbert, a Great Engine of Research*. Austin, University of Texas Press. 9

"is terse to a fault": Gilbert, G. K. (1902). "John Wesley Powell." *Science*: 561-567. 566-567

"he were simply reporting the relative weights": Pyne, S. J. (1980). *Grove Karl Gilbert, a Great Engine of Research*. Austin, University of Texas Press. 103

"To the fact that there were a variety": Ibid. 112

"before he was thrown out": Pyne, S. J. (1999). *How the Canyon Became Grand: A Short History*. New York, Penguin Books. 72

"cultivated the Survey men": Ibid. 73

"Possibly in part for this reason": Worster, D. (2002). *A River Running West: The Life of John Wesley Powell*. New York, Oxford University Press. 322

"whose right to do so [was] virtually prescriptive": Dutton, C. E. (2001). *Tertiary History of the Grand Cañon District*. Tucson, University of Arizona Press. vii

"[Powell's] direction of the Survey": Dutton, C. E. (1879). "The geological history of the Colorado River and Plateaus." *Nature* XIX: 247-252, 272-275. 274-275

"if a full accounting were called for": Dutton, C. E. and J. W. Powell (1885). "Mount Taylor and the Zuni Plateau, New Mexico." *United States Geological Survey Annual Report* 6: 105-198

"If Gilbert was Powell's right hand": Stegner, W. E. (1992). *Beyond the Hundredth Meridian: John Wesley Powell and the Second Opening of the West*. New York, Penguin Books. 158

"Though it seems clear": Ibid. 159

"In the end, politics came between them": Ibid. 330. Pyne, S. J. (1999). *How the Canyon Became Grand: A Short History*. New York, Penguin Books. 85

Chapter 7. The Sublimest Thing

"all around me are interesting geological records": Powell, J. W. (1962). *Exploration of the Colorado River of the West and Its Tributaries*. Cambridge, Harvard University Press. 89

"should be described in blank verse": Dutton, C. E. (1880). *Report on the Geology of the High Plateaus of Utah*. Washington, D.C., U.S. Government Printing Office. 284

"Of Clarence Edward Dutton's three books": Dutton, C. E. (2001). *Tertiary History of the Grand Cañon District*. Tucson, University of Arizona Press. The book actually covers more than just the Tertiary rocks of the area.

"departed from the severe ascetic style": Ibid. viii

"a rich and embroidered nineteenth century traveler's prose": Stegner, W. E. (1992). *Beyond the Hundredth Meridian: John Wesley Powell and the Second Opening of the West*. New York, Penguin Books. 164

"an extraordinary ensemble": Pyne, S. J. (1999). *How the Canyon Became Grand: A Short History*. New York, Penguin Books. 69

"The modern author, pecking away": Ibid. 72

"The lover of nature": Dutton, C. E. (2001). *Tertiary History of the Grand Cañon District*. Tucson, University of Arizona Press. 141-142

"But Dutton had to satisfy only his mentor": Stegner, W. E. (1992). *Beyond the Hundredth Meridian: John Wesley Powell and the Second Opening of the West*. New York, Penguin Books. 170

"*Geology of the High Plateaus of Utah*": Dutton, C. E. (1880). *Report on the Geology of the High Plateaus of Utah*. Washington, D.C., U.S. Government Printing Office.

"distrust[ed] my fitness for the work": Ibid. xv

"many imperfections and . . . falls far short": Ibid. xv

"has a spine like a Stegosaurus": Stegner, W. E. (1992). *Beyond the Hundredth Meridian: John Wesley Powell and the Second Opening of the West*. New York, Penguin Books. 161

"And now the relation of the High Plateaus to the Plateau Province": Dutton, C. E. (1880). *Report on the Geology of the High Plateaus of Utah*. Washington, D.C., U.S. Government Printing Office. 22-23

"A picture of desolation and decay": Ibid. 19

"From the southwest salient of the Markágunt": Ibid. 208-209

"The Aquarius should be described in blank verse": Ibid. 284-286

"at a high pass": Ibid. 297

"the platform of the Kaiparowits Plateau": Ibid. 298

"Print something in an edition as large": Stegner introduction to Dutton, C. E. (2001). *Tertiary History of the Grand Cañon District*. Tucson, University of Arizona Press. vi

"was reduced to typing the entire book": Stegner introduction to Ibid. vii

"the most magnificent picture of the Grand Canyon": Stegner introduction to Ibid. x

"The atlas of **Tertiary History**": Worster, D. (2002). *A River Running West: The Life of John Wesley Powell.* New York, Oxford University Press. 327

"The Grand Canyon challenges both the artist and scientist": Ibid. 329

"Before the observer who stands upon a southern salient": Dutton, C. E. (2001). *Tertiary History of the Grand Cañon District.* Tucson, University of Arizona Press. 26-27

"I shall ask the reader to travel with me": Foster, M. (1994). *Strange Genius: The Life of Ferdinand Vandeveer Hayden.* Niwot, CO, Roberts Rinehart Publishers. 196

"Late in the autumn of 1880 I rode": Dutton, C. E. (2001). *Tertiary History of the Grand Cañon District.* Tucson, University of Arizona Press. 54-55

"We continue to cross hills and valleys": Ibid. 139

"must be cultivated before they can be understood": Ibid. 142

"the sublimest thing on earth": Ibid. 143

"the experience of the sublime": Worster, D. (2002). *A River Running West: The Life of John Wesley Powell.* New York, Oxford University Press. 308

"feel like mere insects crawling along the street of a city": Dutton, C. E. (2001). *Tertiary History of the Grand Cañon District.* Tucson, University of Arizona Press. 86

Chapter 8. Earth's Engine

"the lake shrank away very slowly towards the north": Ibid. 219

"upon the floor of this basin": Ibid. 219

"In no great length of time Ontario would be drained": Dutton, C. E. (1879). "The geological history of the Colorado River and Plateaus." *Nature* XIX: 247-252, 272-275. 250

"would see little but Sphinxes": Ibid. 251

"Of all the changing features of a continent": Ibid. 251-252

"It would be difficult to point out": Ibid. 252

"The structural deformations of the region": Ibid. 252

"the great ocean basins are permanent features": Oreskes, N. (1999). *The Rejection of Continental Drift: Theory and Method in American Earth Science.* New York, Oxford University Press. 208

"One of Willis's last papers": Willis, B. (1944). "Continental drift, Ein Maerchen." *American Journal of Science* 242(9): 509-513

"To be uncertain is to be uncomfortable": Cited in Holmes, A. (1957). "Response to Penrose Medal Citation." *Proceedings of Geological Society of America for 1956*: 75

"star entered the hole": Hoyt, W. G. (1987). *Coon Mountain Controversies: Meteor Crater and the Development of Impact Theory*. Tucson, University of Arizona Press.

"What is harder to understand": Ibid.

"The long shadow that Gilbert cast": Powell, J. L. (2001). *Mysteries of Terra Firma: The Age and Evolution of the Earth*. New York, Free Press. 175-176

"Erosion viewed in one way": Dutton, C. E. (2001). *Tertiary History of the Grand Cañon District*. Tucson, University of Arizona Press. 62

"Those areas which have been uplifted": Dutton, C. E. (1879). "The geological history of the Colorado River and Plateaus." *Nature* XIX: 247-252, 272-275. 251

"few geologists question": Ibid. 251

"the explanation is not quite complete": Ibid. 251

"easy to believe": Ibid. 251

"Dutton wrote several articles": Dutton, C. E. (1931). "On some of the greater problems of physical geology." *Bulletin of the National Research Council* 11(15): 201-211.

"The greatest problems of physical geology": Ibid. 201

"In short, [contraction] could not": Ibid. 202

"For this condition": Ibid. 203

"atomic energy could power": Oreskes, N. (1999). *The Rejection of Continental Drift: Theory and Method in American Earth Science*. New York, Oxford University Press. 51

"the mean rigidity of the subterranean masses": Dutton, C. E. (1931). "On some of the greater problems of physical geology." *Bulletin of the National Research Council* 11(15): 201-211. 205

"if . . . continental blocks really do float": Wegener, A. (1928). *The Origin of Continents and Oceans*. New York, Dover. 45

"In 1967": Oreskes, N. (1999). *The Rejection of Continental Drift: Theory and Method in American Earth Science*. New York, Oxford University Press. Powell, J. L. (2001). *Mysteries of Terra Firma: The Age and Evolution of the Earth*. New York, Free Press.

"great innovations, whether in art or literature": Dutton, C. E. (2001). *Tertiary History of the Grand Cañon District*. Tucson, University of Arizona Press. 141-142

Chapter 9. Where Everything Is Exposed

"Gilbert set out the many advantages": Gilbert, G. K. (1876). "The Colorado Plateau Province as a Field for Geological Study." *American Journal of Science, Third Series* XII: 16-24, 85-103

"its velocity remains the same": Ibid. 99

"its capacity for transportation": Ibid. 99-100

"A stream, which has a supply of debris": Ibid. 100

"from the engineering term": Pyne, S. J. (1980). *Grove Karl Gilbert, a Great Engine of Research*. Austin, University of Texas Press. 28

"The differentiation will proceed": Gilbert, G. K. (1876). "The Colorado Plateau Province as a Field for Geological Study." *American Journal of Science, Third Series* XII: 16-24, 85-103. 100

"Where the bedrock is soft": Ibid. 101

"The Problem of Inconsequent Drainage": Ibid. 101

"Unless the displacements are produced": Ibid. 101

"A large share of the drainage of the Plateaus is not consequent": Ibid. 102

"Emmons was also one of those fired": Emmons, S. F. (1897). "The origin of Green River." *Science* 6: 19-21

"Wallace Stegner titles the chapter": Stegner, W. E. (1992). *Beyond the Hundredth Meridian: John Wesley Powell and the Second Opening of the West*. New York, Penguin Books. 345

"When he was only twelve": Worster, D. (2002). *A River Running West: The Life of John Wesley Powell*. New York, Oxford University Press. 44

"Powell took action": Powell, J. W. (1879). *Report on the Lands of the Arid Region of the United States, With a More Detailed Account of the Lands of Utah*. Harvard and Boston, Massachusetts, The Harvard Common Press.

"usually blond from drought": Reisner, M. (1987). *Cadillac Desert: The American West and Its Disappearing Water*. New York, Penguin Books. 37

"I therefore give it as my firm opinion": Ibid. 37

"Hayden himself endorsed the theory": Foster, M. (1994). *Strange Genius: The Life of Ferdinand Vandeveer Hayden*. Niwot, CO, Roberts Rinehart Publishers. 182

"These men were not crackpots": Reisner, M. (1987). *Cadillac Desert: The American West and Its Disappearing Water*. New York, Penguin Books. 37

"But there wasn't enough water": Ibid.

"It takes a lot more tin": Ibid.

"sneering at natural reality": Ibid. 49

"The change from geographic barbarism": Davis, W. M. (1915). "Biographical memoir of John Wesley Powell, 1834–1902." *Biographical Memoirs*: 11-83. 50

"Of course I have got a great respect": Stegner, W. E. (1992). *Beyond the Hundredth Meridian: John Wesley Powell and the Second Opening of the West*. New York, Penguin Books. 328-331

"The Director was under a misapprehension": Worster, D. (2002). *A River Running West: The Life of John Wesley Powell*. New York, Oxford University Press. 499

"The tight loyalty of the bureau had been cracked": Stegner, W. E. (1992). *Beyond the Hundredth Meridian: John Wesley Powell and the Second Opening of the West*. New York, Penguin Books. 330

"was identical with the program": Ibid. 235

"Scientists wage bitter warfare": Ibid. 324

"kindly and ironic condescension": Ibid. 326

"Little men with big heads": Ibid. 327

"The reduction of the Irrigation Survey": Ibid. 337

"Powell's supporters got the sum back": Ibid. 339

"To add injury": Worster, D. (2002). *A River Running West: The Life of John Wesley Powell*. New York, Oxford University Press. 414

"When the temple came down": Stegner, W. E. (1992). *Beyond the Hundredth Meridian: John Wesley Powell and the Second Opening of the West*. New York, Penguin Books. 342

"more sweeping powers in certain matters": Ibid. 251

"While the fate of the Indian is sealed": Worster, D. (2002). *A River Running West: The Life of John Wesley Powell*. New York, Oxford University Press. 208

"Others have told the tale": Hundley, N. (2001). *The Great Thirst: Californians and Water—A History*. Berkeley, University of California Press. Reisner, M. (1987). *Cadillac Desert: the American West and Its Disappearing Water*. New York, Penguin Books.

"His humane idea": Harris, T. (1991). *Death in the Marsh*. Washington, D. C., Island Press.

"By the time his Geographical Report appeared": Pyne, S. J. (1999). *How the Canyon Became Grand: A Short History*. New York, Penguin Books. 66

"When he died in 1901": Stegner, W. E. (1992). *Beyond the Hundredth Meridian: John Wesley Powell and the Second Opening of the West*. New York, Penguin Books. 344. Wilkins, T. and C. L. Hinkley (1988). *Clarence King: A Biography*. Albuquerque, University of New Mexico Press.

"Let us imagine Major Powell": Stegner, W. E. (1992). *Beyond the Hundredth Meridian: John Wesley Powell and the Second Opening of the West*. New York, Penguin Books. 352, 367

Chapter 10. Antecedence in Doubt

"He began his 1897 article": Powell, J. W. (1875). *Exploration of the Colorado River of the West and Its Tributaries*. Washington, D.C., U.S. Government Printing Office. 166

"on further study": Emmons, S. F. (1897). "The origin of Green River." *Science* 6: 19-21. 20

"Recurring again to the valleys of the Uinta Mountains": Powell, J. W. (1875). *Exploration of the Colorado River of the West and Its Tributaries*. Washington, D.C., U.S. Government Printing Office. 166

"washed either flank of the range": Emmons, S. F. (1897). "The origin of Green River." *Science* 6: 19-21. 20

"What, then, became of the river": Ibid. 20

"In a paper he wrote": Ibid. 21

"a strange caprice seizes it": Dutton, C. E. (1879). "The geological history of the Colorado River and Plateaus." *Nature* XIX: 247-252, 272-275. 252

"determined by the configuration of the surface": Emmons, S. F. (1897). "The origin of Green River." *Science* 6: 19-21. 21

"It would seem proper": Ibid. 21

"Hansen's analysis is worth a moment": Hansen, W. R. (1996). *Dinosaur's Restless Rivers and Craggy Canyon Walls*. Vernal, UT, Dinosaur Nature Association. 10

"So useless has the Survey become": Wilhelms, D. E. (1993). *To a Rocky Moon: A Geologist's History of Lunar Exploration*. Tucson, University of Arizona Press. 8

"look in the face the fact": Chorley, R. J., A. J. Dunn, et al. (1973). *The History of the Study of Landforms or The Development of Geomorphology. Vol. 2, The Life and Work of William Morris Davis*. London; New York, Methuen. 727

"Fifty years afterwards": Ibid. 439

"his 'crack piece'": Ibid. 727

"An 1889 paper in *National Geographic* magazine": Davis, W. M. (1889). "The rivers and valleys of Pennsylvania." *National Geographic*: 183-253

"completed its task": Ibid. 203

"Geographical Cycle of W. M. Davis": Adapted from Encyclopedia Britannica, 2004. www.britannica.com

"every drop of rain that falls": Davis, W. M. (1889). "The rivers and valleys of Pennsylvania." *National Geographic*: 183-253

"In an earlier paper": Davis, W. M. (1888). "Geographic methods in geologic investigation." *National Geographic*: 11-26

"whether or not any river": Davis, W. M. (1889). "The rivers and valleys of Pennsylvania." *National Geographic*: 183-253. 204

"For Gilbert, a graded stream was virtually timeless": Pyne, S. J. (1980). *Grove Karl Gilbert, a Great Engine of Research*. Austin, University of Texas Press. 177

"At the end of the second cycle": Davis, W. M. (1901). "An excursion to the Grand Canyon of the Colorado." *Bulletin of the Museum of Comparative Zoology Harvard University* 38: 107-201. 137

"might now modify in some degree": Ibid. 153

"not seem legitimate": Ibid. 158

"Thus it was not quite antecedent nor quite superposed": Davis, W. M. (1889). "The rivers and valleys of Pennsylvania." *National Geographic*: 183-253. 218

"soon perceive that the earth": Davis, W. M. (1913). "The Grand Canyon of the Colorado." *Journal of Geography*: 310-314. 312

"Compare this vast buried plain": Ibid. 313

"one eternity . . . recognized before another": Ibid. 313

"Some visitors foolishly spend part of their time": Ibid. 314

Chapter 11. The Same River Twice?

"We can say that the modern era": Blackwelder, E. (1934). "Origin of the Colorado River." *Geological Society of America Bulletin* 45(3): 551-566

"Because a river devours its own progeny": In Cox, S. (2003). Grand and Young. *University of Arizona Alumnus*. Winter 2003. www.uagrad.org. Lucchitta makes the same point.

"You cannot step twice into the same river": (1979). *The Oxford Dictionary of Quotations*. Oxford; New York, Oxford University Press. 246

"His publications covered fossils": Krauskopf, K. B. (1976). "Eliot Blackwelder, June 4, 1880–January 14, 1969." *Biographical Memoirs* 48: 83-103

"have become classics in geologic literature": Ibid.

"Rising in the high mountains of Wyoming and Colorado": Blackwelder, E. (1934). "Origin of the Colorado River." *Geological Society of America Bulletin* 45(3): 551-566. 554

"The profile of the Colorado": Ibid. 554

"Doubtless in earlier times": Newberry, J. S. (1861). Geological Report. *Report Upon the Colorado River of the West, Explored in 1857 and 1858 by Joseph C. Ives.* Washington, D.C., U.S. Government Printing Office, U.S. Army, Topographical Corps of Engineers: 154. 19-20

"For hundreds of miles east and west": Blackwelder, E. (1934). "Origin of the Colorado River." *Geological Society of America Bulletin* 45(3): 551-566. 554

"Did the Colorado River exist anywhere in Pliocene time": Ibid. 554

"Yet instead of a network of now-dry tributaries": Ibid. 554

"special circumstance of topography": Ibid. 561

"an association that colored his speech": Rodgers, J. (1982). Chester Ray Longwell. *Biographical Memoirs.* Washington, D.C., National Academy of Sciences. 53: 249-264. 249

"the lure of the unknown": Ibid. 250

"Five received the Penrose Medal": Ibid. 253

"He wrote one paper": Longwell, C. R. (1936). "Geology of the Boulder Reservoir floor." *Geological Society of America Bulletin* 47(9): 1393-1476. Longwell, C. R. (1946). "How old is the Colorado River?" *American Journal of Science* 244(12): 817-835

"no possibility that the river was in its present position": Longwell, C. R. (1946). "How old is the Colorado River?" *American Journal of Science* 244(12): 817-835. 823

"doubtful Pliocene": Ibid. 823

"a Miocene date": Ibid. 823

"not far from *Camelops*": Ibid. 828

"Longwell 'pronounced' this specimen Pleistocene": Ibid. 828

"Thus, there is much that is unknown": Longwell, C. R. (1936). "Geology of the Boulder Reservoir floor." *Geological Society of America Bulletin* 47(9): 1393-1476. 1471

"Longwell noted that upstream": Longwell, C. R. (1946). "How old is the Colorado River?" *American Journal of Science* 244(12): 817-835. 831

"The suggestion of an earlier drainage southward": Ibid. 833-834

"mountain lying down": Powell, J. W. (1875). *Exploration of the Colorado River of the West and Its Tributaries.* Washington, D.C., U.S. Government Printing Office. 187

"He discovered the weird Cambrian fossils": Gould, S. J. (1989). *Wonderful Life: The Burgess Shale and the Nature of History.* New York, Norton.

"Walcott made careful observations": Walcott, C. D. (1890). "Study of a line of displacement in the Grand Canyon of the Colorado, in northern Arizona." *Geological Society of America Bulletin:* 49-64. Walcott, C. D. (1895). "Algonkian rocks of the Grand Canyon of the Colorado." *Journal of Geology:* 312-330

"It is difficult to understand": Walcott, C. D. (1890). "Study of a line of displacement in the Grand Canyon of the Colorado, in northern Arizona." *Geological Society of America Bulletin:* 49-64. 60

"would probably necessitate some change": Ibid. 62

"In two papers in the mid-1940s": Babenroth, D. L. and A. N. Strahler (1945). "Geomorphology and structure of the East Kaibab monocline, Arizona and Utah." *Geological Society of America Bulletin* 56: 107-150. Strahler, A. N. (1948). "Geomorphology and structure of the West Kaibab fault zone and Kaibab, Arizona." *Geological Society of America Bulletin* 59(6): 513-540

"stimulat[ed] it to conquests": Strahler, A. N. (1948). "Geomorphology and structure of the West Kaibab fault zone and Kaibab, Arizona." *Geological Society of America Bulletin* 59(6): 513-540

Chapter 12. Paradox

"We are now in a position to trace the origin": Dutton, C. E. (1879). "The geological history of the Colorado River and Plateaus." *Nature* XIX: 247-252, 272-275

"Charles B. Hunt (1906–1997) wrote the first comprehensive review": Hatheway, A. W. (1993). "Biography of Charles Butler Hunt, geologist." *Bulletin of the Association of Engineering Geologists* 30(2): 139-155

"For close geologic observation try drawing": Hunt, C. B. (1969). "John Wesley Powell, his influence on geology." *Geotimes* 14(5): 16-18

"Hunt's Drawing of the Colorado Plateau": Hunt, C. B. (1969). Geologic history of the Colorado River. *The Colorado River Region and John Wesley Powell.* Reston, VA, U.S. Geological Survey Professional Paper: C59-C130

"In early 1945": Hatheway, A. W. (1993). "Biography of Charles Butler Hunt, geologist." *Bulletin of the Association of Engineering Geologists* 30(2): 139-155. 145

"Hunt wryly noted one passage": Ibid.

"Hunt's first publication on the Grand Canyon": Hunt, C. B. (1946). *Guidebook to the Geology of Utah*. Salt Lake City, Utah Geological Survey.

"The two models disagreed by many millions of years": Longwell, C. R. (1946). "How old is the Colorado River?" *American Journal of Science* 244(12): 817-835

"In 1956, the Survey published": Hunt, C. B. (1956). *Cenozoic Geology of the Colorado Plateau*. Washington, D.C., U.S. Government Printing Office.

"The Muddy Creek Formation in Grand Wash trough": Ibid. 82

"Somewhere between the head and foot": Hunt, C. B. (1969). Geologic history of the Colorado River. *The Colorado River Region and John Wesley Powell*. Reston, VA, U.S. Geological Survey Professional Paper: C59-C130. 113

"it would have been a unique and precocious gully": Hunt, C. B. (1956). *Cenozoic Geology of the Colorado Plateau*. Washington, D.C., U.S. Government Printing Office. 85

"A stable, integrated drainage system cannot behave in this manner": Lucchitta, I. (1984). *Development of Landscape in Northwestern Arizona; the Country of Plateaus and Canyons. Landscapes of Arizona*. T. L. Smiley et al. Lanham, MD, University Press of America. 286

"But all would have agreed": Ibid. 286

"McKee, then regarded as one of the leading geologists": Spamer, E. E. (1999). "Grand vision of Edwin D. McKee." *GSA Today* 9: 18-19

"By the time of the publication of the Symposium report": McKee, E. D., R. F. Wilson, et al. (1967). *Evolution of the Colorado River in Arizona; a Hypothesis Developed at the Symposium on Cenozoic Geology of the Colorado Plateau in Arizona, August 1964*. Flagstaff, AZ, Northern Arizona Society of Science and Art.

"at least two different drainage systems": Ibid. 54

"One sandstone formation": Ibid. 54

"a stream emptying into the Gulf of California": Longwell, C. R. (1936). "Geology of the Boulder Reservoir floor." *Geological Society of America Bulletin* 47(9): 1471

"In only 122 words, summed up": Hunt, C. B. (1968). Review of Evolution of the Colorado River in Arizona. *Geotimes* 39

"Two years later, Hunt laid out his ideas": Hunt, C. B. (1969). Geologic history of the Colorado River. *The Colorado River Region and John Wesley Powell*. Reston, VA, U.S. Geological Survey Professional Paper: C59-C130

Chapter 13. Canyon Makers

"Though he used more sophisticated methods": Leopold, L. B. (1969). *The Colorado River Region and John Wesley Powell*. Reston, VA, U.S. Geological Survey Professional Paper. "The rapids and the pools, Grand Canyon." 131-145. 131

"But the pièce de résistance": Hunt, C. B. (1969). Geologic history of the Colorado River. Reston, VA, U.S. Geological Survey Professional Paper: C59-C130

"Grand Canyon—Grand Problem": Ibid. 113

"The relationship between channel segments": McKee, E. D., R. F. Wilson, et al. (1967). *Evolution of the Colorado River in Arizona; a Hypothesis Developed at the Symposium on Cenozoic Geology of the Colorado Plateau in Arizona, August 1964*. Flagstaff, AZ, Northern Arizona Society of Science and Art. 6-9

"The following three sketch maps": Lucchitta, I. (1984). *Development of Landscape in Northwestern Arizona; the Country of Plateaus and Canyons. Landscapes of Arizona*. T. L. Smiley et al. Lanham, MD, University Press of America.

"put the Hualapais in front": Ives, J. C. (1861). *Report Upon the Colorado River of the West, Explored in 1857 and 1858 by Lieutenant Joseph C. Ives . . . Under the Direction of the Office of Explorations and Surveys*. Washington, D.C., U.S. Government Printing Office. 99

"appears to be the youngest formation": Longwell, C. R. (1936). "Geology of the Boulder Reservoir floor." *Geological Society of America Bulletin* 47(9): 1393-1440

"What was the source of so much water?": Hunt, C. B. (1969). Geologic history of the Colorado River. *The Colorado River Region and John Wesley Powell*. Reston, VA, U.S. Geological Survey Professional Paper: C59-C130. 115

"These porous rocks": Hunt, C. B. (1974). *Grand Canyon and the Colorado River, Their Geologic History*. Flagstaff, AZ. 137

"As a body of water, the Colorado River is small": Hunt, C. B. (1969). Geologic history of the Colorado River. *The Colorado River Region and John Wesley Powell*. Reston, VA, U. S. Geological Survey Professional Paper: C59-C130. 127

"enthusiastic and foolish": Lucchitta, I. (2002). Letters from the Grand Canyon: Piracy and Capture Carve the Grand Canyon: Part A. *Boatman's Quarterly Review*. 15. www.gcrg.org

"The hypothetical river downstream": Ibid.

"In 1988 the Museum of Northern Arizona published *Canyon Maker*": Lucchitta, I. and Museum of Northern Arizona. (1988). *Canyon Maker: A Geological History of the Colorado River*. Flagstaff, AZ, Museum of Northern Arizona. Unfortunately, out of print.

"His theory has changed but little": Lucchitta, I. (2003). History of the Grand Canyon and the Colorado River. *Grand Canyon Geology*. S. S. Beus and M. Morales. New York, Oxford University Press. 260-274

"Scattered here and there on the Hualapai Plateau": Young, R. A. and W. J. Brennan (1974). "Peach Springs Tuff; Its Bearing on Structural Evolution of the Colorado Plateau and Development of Cenozoic Drainage in Mohave County, Arizona." *Geological Society of America Bulletin* 85(1): 83-90

"short, steep, immature": Lucchitta, I. and Museum of Northern Arizona. (1988). *Canyon Maker: A Geological History of the Colorado River*. Flagstaff, AZ, Museum of Northern Arizona. 16

"gravels of probable river origin": Lucchitta, I. (2003). History of the Grand Canyon and the Colorado River. *Grand Canyon Geology*. S. S. Beus and M. Morales. New York, Oxford University Press: 260-274. 263

"Geologist George Billingsley mapped these same plateaus": Billingsley, G. H. (2003). Personal communication.

"In 1998, geologists used the argon-argon method": Spencer, J. E., L. Peters, et al. (1998). "6 Ma 40-Ar/39-Ar date from the Hualapai Limestone and implications for the age of the Bouse Formation and Colorado River." *Geological Society of America, Rocky Mountain Section, 50th Annual Meeting*. Geological Society of America (GSA). 30: 37

"extremely sparse material of probably Colorado Plateau derivation": Buising, A. V. (1990). "The Bouse Formation and bracketing units, southeastern California and western Arizona: Implications for the evolution of the proto-Gulf of California and the lower Colorado River." *Journal of Geophysical Research*, 95: 20111-20132

"Fortunately, interbedded with the sedimentary layers": Spencer, J. E., L. Peters, et al. (1998). "6 Ma 40-Ar/39-Ar date from the Hualapai Limestone and implications for the age of the Bouse Formation and Colorado River." *Geological Society of America, Rocky Mountain Section, 50th Annual Meeting*. Geological Society of America (GSA). 30: 37

"American Southwest and the Gulf of California": After Schmidt, N. (1990). "Plate Tectonics and the California Region." *Arizona Geology* 20(2): 1-4

Chapter 14. Lazarus and Lakes

"Scores of geologists had put in years of work": www.grandcanyonbiblio.org/

"In June 2000, seventy-seven specialists convened": Young, R. and E. E. Spamer, Eds. (2004). *Colorado River: Origin and Evolution: Proceedings of the Symposium Held at Grand Canyon National Park in June 2000.* Grand Canyon, AZ, Grand Canyon Association.

"the Symposium report": Ibid.

"At last the movements which began at the commencement of Tertiary time": Powell, J. W. (1876). *Report on the Geology of the Eastern Portion of the Uinta Mountains and a Region of Country Adjacent Thereto.* Washington, D.C., U.S. Government Printing Office. 35

"Glen Canyon died": Hamilton, J. (2001). Passages. *Sierra Magazine.* January/February. www.sierraclub.org

"Doubtless in earlier times": Newberry, J. S. (1861). Geological Report. *Report Upon the Colorado River of the West, Explored in 1857 and 1858 by Joseph C. Ives.* Washington, D.C., U.S. Government Printing Office, U.S. Army, Topographical Corps of Engineers: 154. 19-20

"even the few persistently flowing streams dwindle": Blackwelder, E. (1934). "Origin of the Colorado River." *Geological Society of America Bulletin* 45(3): 551-566. 560

"resembled somewhat its present self": Ibid. 562

"haphazard and accidental development": Ibid. 564

"more than 99 percent of the lake clays have eroded": Meek, N. and J. Douglass (2004). "Lake-Overflow: An alternative hypothesis for Grand Canyon incision and development of the Colorado River." R. Young and E. E. Spamer. *Colorado River: Origin and Evolution: Proceedings of the Symposium Held at Grand Canyon National Park in June 2000.* Grand Canyon, Arizona, Grand Canyon Association. 12: 199-206. 200

"But at the June 2000 Symposium": Dallegge, T. A., M. Ort, et al. (2000). "Age Constraints and Depositional Basin Morphology for the Mid-Pliocene Bidahochi Formation, Northeastern Arizona." R. Young and E. E. Spamer. *Colorado River: Origin and Evolution: Proceedings of the Symposium held at Grand Canyon National Park in June 2000.* Grand Canyon, AZ, Grand Canyon Association. 47-53

"An important scientific innovation rarely makes its way": Planck, M. (1949). *Scientific Autobiography.* New York, Philosophical Library.

Chapter 15. Molten Rock, Melted Snow

"If he can only study geology he will be happy": Worster, D. (2002). *A River Running West: The Life of John Wesley Powell.* New York, Oxford University Press. 184

"Great quantities of cooled lava": Powell, J. W. (1875). *Exploration of the Colorado River of the West and Its Tributaries*. Washington, D.C., U.S. Government Printing Office. 94-95

"Some flows were extruded on the Uinkaret Plateau": Hamblin, W. K. (2003). Late Cenozoic Lava Dams in the Western Grand Canyon. *Grand Canyon Geology*. S. S. Beus and M. Morales. New York, Oxford University Press. 313

"One geologist estimates": Stiles, E. (July 18, 2002). "Is the Grand Canyon a geologic infant?" www.uanews.opi.arizona.edu

"Hamblin concluded": Hamblin, W. K. and L. Hamblin (1997). "Fire and Water: The Making of the Grand Canyon." *Natural History* (September): 35-41. 38

"But unfortunately": Fenton, C., R. Poreda, et al. (2004). "Geochemical discrimination of five Pleistocene lava-dam outburst-flood deposits, western Grand Canyon, Arizona." *Journal of Geology* 112(No. 1): 91-110

"Geologist Robert Webb": Stiles, E. (July 18, 2002). Ibid.

"What had appeared to be lake sediment": Kaufman, D. S., G. O'Brian, et al. (2002). "Late Quaternary spring-fed deposits of the Grand Canyon and their implication for deep lava-dammed lakes." *Quaternary Research* 58: 329-340

"the capacity to erode through any rock": Hamblin, W. K. (2003). Late Cenozoic Lava Dams in the Western Grand Canyon. *Grand Canyon Geology*. S. S. Beus and M. Morales. New York, Oxford University Press. 344

"the process of slope retreat did not enlarge the canyon": Ibid. 345

"Sometimes none do, and the scientist has to give up": Dalrymple, G. B. (1991). *The Age of the Earth*. Stanford, CA, Stanford University Press. Faure, G. (1986). *Principles of Isotope Geology*. New York, Wiley. Powell, J. L. (2001). *Mysteries of Terra Firma: The Age and Evolution of the Earth*. New York, Free Press.

"The strongest evidence of the reliability of the potassium-argon method": See discussion in Powell, J. L. (2001). *Mysteries of Terra Firma: The Age and Evolution of the Earth*. New York, Free Press.

"A group of researchers addressed the first question": Renne, P. R., W. D. Sharp, et al. (1997). "(super 40) Ar/ (super 39) Ar dating into the historical realm; calibration against Pliny the Younger." *Science* 277 (5330): 1279-1280

"The following table": Dallegge, T. A., M. Ort, et al. (2000). "Age Constraints and Depositional Basin Morphology for the Mid-Pliocene Bidahochi Formation, Northeastern Arizona." R. Young and E. E. Spamer. *Colorado River: Origin and Evolution, Proceedings of the Symposium held at Grand Canyon National Park in June 2000*. Grand Canyon, AZ, Grand Canyon Association. 49

"Several authors have dated": McIntosh, W. C., L. Peters, et al. (2002). New 40Ar-39Ar dates on basalts in Grand Canyon; constraints on rates of Quaternary river incision and slip on the Toroweap Fault and implications for lava dams. *Geological Society of America, Rocky Mountain Section, 54th Annual Meeting*, Geological Society of America (GSA). 34: 61. Lucchitta, I., G. H. Curtis, et al. (2000). "Cyclic aggradation and downcutting, fluvial response to volcanic activity, and calibration of soil-carbonate stages in the western Grand Canyon, Arizona." *Quaternary Research* 53(1): 23-33

"The measured incision rate": Pederson, J. L., K. Karlstrom, et al. (2002). "Differential incision of the Grand Canyon related to Quaternary faulting-constraints from U-series and Ar/Ar dating." *Geology* 30(8): 739-742

"some canyon cutting": Ibid. 742

Chapter 16. What Caused the Grand Canyon?

"Most of the drop occurred in the western Grand Canyon": Lucchitta, I. (2002). "Letters from the Grand Canyon: Piracy and Capture Carve the Grand Canyon: Part A." *Boatman's Quarterly Review*. 15. www.gcrg.org

"All geologists appear to have assumed": Blackwelder, E. (1934). "Origin of the Colorado River." *Geological Society of America Bulletin* 45(3): 551-566. 556

"At least two different drainage systems": McKee, E. D., R. F. Wilson, et al. (1967). *Evolution of the Colorado River in Arizona; a Hypothesis Developed at the Symposium on Cenozoic Geology of the Colorado Plateau in Arizona, August 1964*. Flagstaff, AZ, Northern Arizona Society of Science and Art. 54

"The difficulties of interpreting the geologic history": Hunt, C. B. (1969). Geologic history of the Colorado River. *The Colorado River Region and John Wesley Powell*. Reston, VA, U.S. Geological Survey Professional Paper: C59-C130

"The presence of a well-integrated": Young, R. and E. E. Spamer, Eds. (2004). *Colorado River: Origin and Evolution: Proceedings of the Symposium Held at Grand Canyon National Park in June 2000*. Grand Canyon, AZ, Grand Canyon Association. 2

"Some scientists are convinced": Ward, P. D. and D. Brownlee (2000). *Rare Earth: Why Complex Life is Uncommon in the Universe*. New York, Springer.

"It is intuitively more satisfying": Hallam, A. (1981). "The end-Triassic bivalve extinction event." *Palaeogeography, Palaeoclimatology, Palaeoecology* 35(1): 1-44

"Geologic Cross Section of the Grand Canyon": Potochnik, A. R. and S. J. Reynolds (2003). *Side Canyons of the Colorado River in Grand Canyon. Grand Canyon Geology*. S. S. Beus and M. Morales. New York, Oxford University Press. 391-406

Key Terms and Places

"**Key Terms and Places**": www.grandcanyontreks.org An excellent source of information about the many names of Colorado Plateau features.

"**Pyne attributes the term to Gilbert**": Pyne, S. J. (1980). *Grove Karl Gilbert, a Great Engine of Research*. Austin, University of Texas Press. 52

"**Grand Canyon: Powell instructed Dellenbaugh to use this name**": Worster, D. (2002). *A River Running West: The Life of John Wesley Powell*. New York, Oxford University Press. 299

Acknowledgments

Many friends tried their best to help me get both the science and the writing right. David Morrison, Joel Pederson, and Richard Young gave far more time than I could have expected. George Billingsley, Paul Knauth, Norman Meek, Donald Reich, and Robert Scarborough made many helpful contributions. The support of my agent, John Thornton, was the *sine qua non* of the project. The confidence of my editor, Stephen Morrow, inspired me to keep going. Janet Fong drafted the illustrations; the wonderful Kim Llewellyn designed the book from cover to cover. As always, my wife Joan and daughter Joanna were patient and supportive. I dedicate this book to Amelia Rose, Kemar, Sophie Jo, and Yawaske.

Index

Figure Credits